BULKING OF ACTIVATED SLUDGE:
Preventative and Remedial Methods

Other titles published in collaboration between

ELLIS HORWOOD LIMITED and
THE WATER RESEARCH CENTRE

RIVER POLLUTION CONTROL
Edited by M. J. Stiff, WRC

BIOLOGICAL FLUIDISED BED TREATMENT OF WATER AND WASTEWATER
Edited by P. F. Cooper, WRC, and B. Atkinson, UMIST

AQUALINE THESAURUS
Compiled by J. G. Smith and edited by P. J. Russell, WRC

WATER RESEARCH TOPICS Volume 1
Edited by I. M. Lamont, WRC

BULKING OF ACTIVATED SLUDGE: Preventative and Remedial Methods

Editors:
B. CHAMBERS and E. J. TOMLINSON
Process Engineering
Water Research Centre
Stevenage Laboratory
Hertfordshire

Published by
ELLIS HORWOOD LIMITED
Publishers · Chichester

for the
WATER RESEARCH CENTRE

First published in 1982 by

ELLIS HORWOOD LIMITED

Market Cross House, Cooper Street, Chichester, West Sussex, PO19 1EB, England

The publisher's colophon is reproduced from James Gillison's drawing of the ancient Market Cross, Chichester.

Distributors:

Australia, New Zealand, South-east Asia:
Jacaranda-Wiley Ltd., Jacaranda Press,
JOHN WILEY & SONS INC.,
G.P.O. Box 859, Brisbane, Queensland 40001, Australia

Canada:
JOHN WILEY & SONS CANADA LIMITED
22 Worcester Road, Rexdale, Ontario, Canada.

Europe, Africa:
JOHN WILEY & SONS LIMITED
Baffins Lane, Chichester, West Sussex, England.

North and South America and the rest of the world:
Halsted Press: a division of
JOHN WILEY & SONS
605 Third Avenue, New York, N.Y. 10016, U.S.A.

British Library Cataloguing in Publication Data
Bulking of activated sludge:f Preventative and Remedial Methods
1. Sewage – Activated Sludge processes
I. Chambers, B.   II. Tomlinson, E. J.
628.3'54   TD756

Library of Congress Card No. 81-20079 AACR2

ISBN 0-85312-350-0 (Ellis Horwood Limited, Publishers)
ISBN 0-470-27299-6 (Halsted Press)

Typeset in Press Roman by Ellis Horwood Ltd.
Printed in Great Britain by R. J. Acford, Chichester.

Table of Contents

PART 2 MICROBIOLOGICAL ASPECTS – THE STATE OF THE ART

PART 3 PROGRESS THROUGH LABORATORY AND PILOT SCALE STUDIES

Editors' Preface

This book is derived from the Water Research Centre conference entitled 'Bulking of Activated Sludges: Prevention or Cure' held in Cambridge in April 1981. The activated sludge process has been used widely in waste water treatment since its initial development nearly seventy years ago. The advantages of the process in comparison with biological filtration were soon recognized and so too was the major disadvantage, namely the operating problems resulting from poor sludge settleability or 'bulking'. Despite the vast literature available on the causes and control of sludge bulking, poor settleability is still estimated to be the cause of operating problems at about half of the full-scale plants installed worldwide. The WRC has been performing research on the phenomenon of sludge bulking for many years and it was felt that the time was right for a conference devoted to all aspects of the problem. Accordingly, our aim was to assemble all researchers working in this area and attempt to produce definitive statements which would result in more confident design and operation of activated-sludge systems.

The conference covered all aspects of the sludge bulking problem and many authors presented the results of original research. Considerable attention was devoted to the microbiology of the activated-sludge process, in particular, the role of filamentous micro-organisms in causing poor sludge settleability. Other papers presented the results of pilot-scale experimental work and investigations performed on full-scale plant. Many of the papers generated lively discussions between authors and delegates and the discussion sessions are recorded here in detail at the end of each chapter. A number of poster papers were displayed at the conference and these have been included in the book as an Appendix.

We believe that this book fulfils our aim of providing a definitive and up to data statement on all aspects of sludge bulking and hence should prove valuable to all those in the Water Industry who are involved in the design and operation of activated-sludge plants.

The Water Research Centre expresses thanks to all those who contributed to the conference and the publication of this book especially authors, session chairmen and Mr. A. G. Boon who gave the closing address. Thanks are also due to Mrs. A. J. Vincent and Mr. P. F. Cooper for acting as rapporteurs of the discussion sessions and to Miss M. Kennedy-Ping for preparing the book for publication.

B. Chambers
E. J. Tomlinson

Stevenage, November 1981

Directory of Contributors

		Chapter
Bode, H.	Wirtschaft Technische Universitat Hannover, Welfengarten 1, 3000 Hannover 1, West Germany	11
Chambers, Dr. B.	WRC Process Engineering, Stevenage Laboratory, Elder Way, Stevenage, Hertfordshire. SG1 1TH	10 Appendix I (v)
Chudoba, J.	Department of Water Technology and Environmental Engineering, Institute of Chemical Technology, 166 28 Prague 6, Czechoslovakia.	7
Critchard, D. J.	Area Chemist, Thames Water Authority Lambourne Division, 24 Barossa Road, Camberley, Surrey.	13
Deinema, Dr. M. H.	Department of Microbiology, Agricultural University, Hesselink van Suchtelenweg 4/6703, Wageningen, The Netherlands.	5
Dohányos, M.	Department of Water Technology and Environmental Engineering, Institute of Chemical Technology, 166 28 Prague 6, Czechoslovakia.	7
Donker, H. J. G. W.	Department of Water Pollution Control, Agricultural University, Biotechnion, De Dreijen 12/6703 BC Wageningen, The Netherlands.	9
Eikelboom, Ir. D. H.	Research Institute for Environmental Hygiene, TNO, Postbus 214, Schoemakerstraat 97, 2600 AE Delft, The Netherlands.	3 and 6

Grau, Professor Dr. P.	Head, Department of Water Technology and Environmental Engineering, Institute of Chemical Technology, 166 28 Prague 6, Czechoslovakia.	7
Green, Dr. G. P.	Biologist, Regional Biology Unit, Wessex Water Authority, Avon and Dorset Division, 2 Nuffield Road, Poole, Dorset.	4
Hanbury, Dr. M. J.	Thames Water Authority, New River Head, Rosebery Avenue, London EC1R 4TP	13
Houtmeyers, Dr. J.	Forest Biology Centre, Bokrijk-Genk, Belgium.	8
Humphries, M. D.	Assistant Chemist, Wessex Water Authority, Bristol Avon Division, Chippenham District Office, 38 St. Paul Street, Chippenham, Wiltshire. SN15 1LY	Appendix I (iv)
Hurley, B. J. E.	Principal Research Scientist, Thames Water Authority, Directorate of Scientific Services, New River Head, Rosebery Avenue, London EC1R 4TP	12
Ijwema, T. S. J.	Department of Water Pollution Control, Agricultural University, Biotechnion, De Dreijen 12/6703 BC Wageningen, The Netherlands.	9
Jenkins, Professor D.	University of California, Berkeley, College of Engineering, Sanitary Engineering Research Laboratory, Richmond Field Station, 1301 South 46th Street, Richmond, California 94804, USA	11
Johnstone, Dr. D. W. M.	Divisional Scientist, Thames Water Authority, Cotswold Division, 17 Bath Road, Swindon, Wiltshire. SN1 4AT	13
Neethling, Dr. J. B.	University of California, Berkeley, College of Engineering, Sanitary Engineering Research Laboratory, Richmond Field Station, 1301 South 46th Street, Richmond, California 94804, USA	11
Price, G. J.	Senior Chemist, Wessex Water Authority, Bristol Avon Division, Chippen-	Appendix I (iii)

	ham District Office, 38 St. Paul Street, Chippenham, Wiltshire. SN15 1LY.	
Rachwal, A. J.	Research Scientist, Thames Water Authority, Central Research Laboratory, Northumberland House, Mogden Sewage Treatment Works, Mogden Lane, Isleworth, Middlesex.	13
Rader, W. E.	Shell Development Company, Westhollow Research Center, PO Box 1380, Houston, Texas 77001, USA	Appendix I (iii)
Rensink, Ir. J. H.	Department of Water Pollution Control, Agricultural University, Biotechnion, De Dreijen 12/6703 BC Wageningen, The Netherlands.	9
Richard, Dr. M. G.	University of California, Berkeley, College of Engineering, Sanitary Engineering Research Laboratory, Richmond Field Station, 1301 South 46th Street, Richmond, California 94804, USA.	11
Salanitro, J. P.	Microbiologist, Shell Development Company, Westhollow Research Center, PO Box 1380, Houston, Texas 77001, USA	Appendix I (ii)
Slijkhuis, Ir. H.	Department of Microbiology, Agricultural University, Hesselink van Suchtelenweg 4/6703, Wageningen, The Netherlands.	5
Tomlinson, E. J.	WRC Process Engineering, Stevenage Laboratory, Elder Way, Stevenage, Hertfordshire. SG1 1TH.	1 Appendix I (v)
Van den Eynde, E.	University of Louvain, Laboratory of Industrial Microbiology and Biochemistry, Kard. Mercierlaan 92, Heverlee B3030, Belgium.	8
Verachtert, Professor H.	University of Louvain, Laboratory of Industrial Microbiology and Biochemistry, Kard. Mercierlaan 92, Heverlee B3030, Belgium	8
Wagner, F.	Institute of Water Conservation, Water Purification and Industrial Waste, University of Stuttgart, Bandtale 1, 7000 Stuttgart 80, West Germany	2

Walker, A. P.	North West Water Authority, Microbiology Laboratory, Stonebridge Lane, Liverpool L10 5HA	Appendix I (i)
Waller, C. B.	Divisional Scientist, Thames Water Authority, Vales Division, Denton House, Iffley Turn, Oxford, OX4 4HJ	12

PART 1: THE PROBLEM

Chapter 1

The emergence of the bulking problem and the current situation in the UK

E. J. Tomlinson, Water Research Centre, Stevenage Laboratory

1.1 INTRODUCTION

The term bulking has been used for many years to describe problems associated with the separation of activated sludge. A precise definition of the term is difficult since in the past it has been used to describe a variety of problems. This has resulted in considerable confusion for those trying to investigate the phenomenon. For the purposes of this chapter, bulking is defined as the phenomenon which occurs in activated-sludge plants when the sludge occupies an excessive volume and does not settle readily, so that in extreme cases the effluent from secondary settlement tanks contains an excessive amount of suspended matter. It is important to stress that this definition does not embrace the phenomena of deflocculation or rising sludge and this is intentional.

There is no indication in the literature to suggest that problems of poor settleability occurred when the first experiments with activated sludge were performed in the early part of this century. However, by the early 1920s a number of problems had been reported and from that time it became apparent that these problems could be serious and were widespread. Since then the results of many experiments and operational experiences have been reported and these have clearly led to a better understanding of the activated-sludge process. However, the problem still exists and it is doubtful whether any significant reduction has been achieved in the extent of the problem.

The effects of bulking on the quality of final effluent from an activated-sludge plant will depend on the seriousness of the problem and the design of the secondary settlement tanks. Where settleability is poor, but not excessively so, and the design of secondary settlement tanks is reasonably conservative, it is possible for effluent quality to be improved. The reason for this seems to be that under such circumstances a deep sludge blanket forms in the settlement tank and the latter then acts as a blanket clarifier. Filtration of suspended material occurs in the blanket and this results in a final effluent with a lower suspended-solids content and a proportionately reduced biochemical oxygen demand. However, the diurnal variation in waste-water flow experienced by most treatment plants means that this mode of operation is highly unstable and cannot normally be considered as a reliable means of improving effluent quality.

In cases where the solids-handling capacity of the secondary settlement tank is exceeded, serious loss of activated sludge from the system occurs. In the first instance this will result in an increase in the biochemical oxygen demand of the final effluent which is proportional to the amount of solids lost. If the loss continues, the reduction in mixed-liquor suspended-solids concentration in the aeration tank will bring about a reduction in the sludge age for the plant. The ensuing loss of nitrification will be responsible for further deterioration in effluent quality. In the extreme cases of continued loss of solids, a reduction in carbonaceous oxidation will take place and ultimately a complete loss of treatment may occur.

1.2 THE SITUATION IN THE UK

The most recent survey of the situation in the UK was that carried out by the Water Research Centre and reported in 1976 [1]. The survey had two primary aims. Firstly, it established the extent of the problem and secondly it attempted to identify some of the causes.

The initial results showed that there was no single, simple criterion which could be used to establish the occurrence of bulking. Two criteria were eventually adopted and these were based upon the yearly average Sludge Volume Index (SVI) value of the sludge and the occurrence of excessive loss of solids over the weirs of final settlement tanks. Out of 65 plants studied, 27 had yearly average SVI values in excess of 200 ml/g. Thirty-four of the plants experienced excessive loss of solids. The 27 plants with high values of SVI were not all amongst the 34 which experienced loss of solids. This suggested that some plants were able to operate satisfactorily in spite of high SVI values. A closer study of the data showed that in such cases loss of solids was avoided either because abnormally high settlement tank capacity existed or because operation at very low concentrations of mixed-liquor suspended solids was practised. In all, 41 plants (63 per cent) had either serious losses of solids or an average SVI value greater than 200 ml/g, or both. Similar conditions exist in Europe and North America and subsequent chapters will refer to these. It is unlikely that the situation in the UK has changed significantly since 1976, primarily because the solutions to the problems are often costly and therefore unattractive.

1.3 APPROACHES TO THE PROBLEM

Investigations into the phenomenon of bulking have been undertaken for the past 60 years. Initially, investigators attempted to analyse plant design, operating conditions and waste-water characteristics and relate these to the settleability of sludge. Accordingly, the occurrence of bulking was, at different times, associated with overloading, underloading, over-aeration, under-aeration, short-circuiting, nutrient imbalance, high pH-value, low pH-value, high temperature, sewage septicity and many other identifiable conditions. This approach yielded solutions to individual problems though in many cases there was no possible remedial action which could be taken even when the condition thought to be responsible for bulking had been identified. From this situation resulted the publication of much conflicting information and advice and it soon became apparent that the solutions, though applicable at particular locations, could not be adoped universally.

The settleability of activated sludge is, of course, affected by the compostion and characteristics of its microflora. This fact was recognized and formed the basis of later, more scientifically rigorous, approaches. Attempts were made to relate sludge settleability to fundamental observations of micro-organism type and surface chemistry characteristics. Unfortunately, two distinct schools of thought developed, those who believed that bulking resulted from excessive growth of certain filamentous micro-organisms, and those who believed that it resulted from the presence or absence of certain physico-chemical conditions on the surfaces of the micro-organisms. Judging by the positive results obtained by both schools, it must be concluded that both theories are relevant. Unfortunately, neither approach has led directly to a solution of the problem. The reason for this is that even when the relationships between fundamental characteristics and settleability have been established it is still necessary to determine how these fundamental characteristics are affected by design, operating conditions and waste-water characteristics. It is in this area that progress has been very slow and, as subsequent chapters will show, recent and current investigations have tended to seek empirically-based solutions rather than ones based on fundamental scientific principles.

1.4 DEVELOPMENT OF THE ACTIVATED-SLUDGE PROCESS

It is appropriate to consider the development of the process over the past 65 years, since this throws considerable light on the bulking problem. Initial experiments, performed in Manchester, employed the so-called 'fill-and-draw' method of treatment. Typically this consisted of 'filling' a tank with sewage over a one-hour period, aerating the contents of the tank for three hours and then allowing quiescent settlement for two hours. Finally, the clear liquid was 'drawn' from the upper part of the tank over a one-hour period, leaving activated sludge in the bottom of the tank ready for the next batchwise cycle. No problems with poor settleability were reported. However, it was appreciated very quickly that a continuous-flow process was desirable and subsequent experiments employed separate tanks for aeration and settlement. Sewage was fed continuously to the aeration tank and the concentration of activated sludge in this tank was maintained by recycling sludge from the bottom of the settlement tank. Many modifications of the original theme developed, including contact-stabilization, extended aeration and replacement of coarse-bubble aeration by diffused-air or mechanical aeration systems. Aeration tank and settlement tank geometry varied enormously, and within 25 years there were almost as many different designs as there were separate activated-sludge plants. The problems associated with poor settleability were, by then, well known to many plant operators. In 1943 a survey of 21 diffused-air activated-sludge plants in North America indicated that 12 of these experienced problems either continuously or intermittently [2]. Two years later, a further survey [3] of 29 mechanically-aerated plants showed that only eight of these had experienced little or no trouble due to bulking.

Many reasons have been put forward for the occurrence of bulking and some of these are referred to in subsequent chapters. However, it is the contention of this author that many of the problems have arisen because of fundamental design differences between original plant and that used since. These differences can be elucidated as follows:

In the original 'fill-and-draw' tests, a relatively large volume of sewage was mixed with a small volume of sludge in a relatively short time. Thus, micro-organisms initially experienced a high concentration of carbonaceous substrate and this gradually reduced as oxidation proceeded over a period of time. The aeration tanks of the original continuous-flow plants had a high length-to-breadth ratio [4] and this ensured that fully-oxidized effluent near the aeration-tank outlet did not dilute sewage as it entered with recycled activated sludge at the inlet. Therefore, these plants also provided a high initial concentration of carbonaceous substrate to micro-organisms in the the same way that the 'fill-and-draw' plants did.

The situation in the aeration tanks of these continuous-flow plants should be described more correctly in terms of dispersion or longitudinal mixing. Tanks with high length-to-breadth ratios exhibit very little longitudinal mixing or dispersion characteristics and are commonly referred to as 'plug-flow' configurations.

There is now considerable evidence that many sludge bulking problems can be associated with the departure from the conditions provided by 'fill-and-draw' plants and by the early continuous-flow plants. This departure has come about because of the more widespread use of so-called 'completely-mixed' aeration tanks, or more correctly, aeration tanks with a high degree of dispersion or longitudinal mixing. This type of mixing characteristic ensures that, as sewage and recycled sludge enter the aeration tank, they are immediately dispersed throughout the existing contents. Accordingly, the micro-organisms experience only one concentration of carbonaceous substrate, that being the very low concentration remaining in the treated sewage.

In some cases this is a distinct advantage, for example, where toxic or inhibitory trade effluents are present in sewage it is desirable to dilute them immediately they enter the aeration tank. This is probably one of the main reasons for proliferation of the 'completely-mixed' systems. Another reason for the increase in use during the 1950s was that vertical-shaft mechanical aeration devices were popular at that time and these were better suited to 'completely-mixed' aeration tanks.

Considerable evidence now exists to support the view that poor settleability of activated sludge can be associated with the mixing characteristics of the aeration tank. As long ago as 1932 it was suggested [5, 6] that 'short-circuiting' in aeration tanks could cause bulking and that the installation of baffles would overcome the problem. The installation of baffles stops 'short-circuiting' by reducing the degree of longitudinal mixing and this, of course, helps to create conditions closer to those employed in the early activated-sludge plants.

More recently Pasveer [7], who was himself responsible for the development of the now well-known oxidation-ditch modification to the activated-sludge process, provided further evidence for the argument. Early versions of the oxidation-ditch employed the 'fill-and-draw' technique. Pasveer noticed that, as these plants were converted to 'completely-mixed', 'continuous-flow' operation, poor settleability became common. Subsequent large-scale experiments [8] confirmed these observations and also showed that even when a 'fill-and-draw' plant is employed, poor sludge settleability can occur if the 'fill' period is long and the

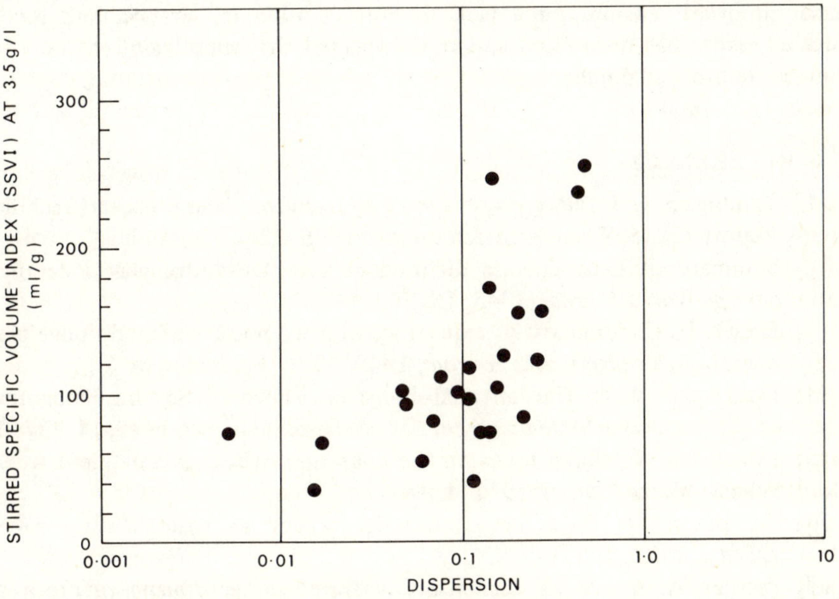

Fig. 1.1 – Relation between settleability and dispersion

tank is aerated during this long 'fill' period. He concluded that these latter conditions also resulted in low carbonaceous substrate loadings similar to those provided by 'completely-mixed' systems.

In the United Kingdom, the mixing characteristics of more than 20 full-scale activated-sludge plants [9] have been quantified by the use of 'stimulus-response' techniques [10] and the values correlated with sludge settleability. The results, shown in Fig. 1.1, again confirm that 'completely-mixed' aeration-tank configurations result in sludges with poor settling characteristics.

In addition to the information referred to above, there is also extensive evidence from laboratory-scale experiments to support the argument [11–13] and further examples will be described in later chapters.

1.5 CONCLUSIONS

Poor settleability of activated sludge was not a problem encountered during early development of the process. Many modifications and developments of the original process have been adopted by designers over the past 50 years and many different waste waters, sometimes of indefinite compostion, have been treated. There has been insufficient consideration given to the effects of changes in design, operating conditions and waste-water characteristics upon the settleability of activated sludge and often full-scale plants have been designed without any laboratory or pilot-scale investigations. This situation is unknown in other process industries and should not continue in the fields of waste-water treatment. One important example, the lack of consideration of aeration-tank mixing characteristics has been cited and it is expected that subsequent chapters will illuminate other examples.

1.6 REFERENCES

[1] Tomlinson, E. J. *Bulking – A Survey of Activated-Sludge Plants.* Technical Report TR 35, Water Research Centre, 1976, 32 pp.

[2] Summary of experience in diffused-air activated-sludge plant operation. *Sewage Works Journal,* 1943, **15**, 909–940.

[3] Dreier, D. E. Summary of experience in mechanical activated-sludge plant operation. *Sewage Works Journal,* 1945, **17**, 101–126.

[4] Duckworth, W. H. The activated-sludge experiments at Salford. *Proceedings of the Association of Managers of Sewage Disposal Works,* 1915, **14**, 73–82.

[5] Donaldson, W. Some notes on the operation of sewage treatment works. *Sewage Works Journal,* 1932, **4**, 48–59.

[6] Donaldson, W. Use of activated-sludge process increasing. *Sewage Works Journal,* 1932, **4**, 567–568.

[7] Pasveer, A. A case of filamentous activated sludge. *Journal of the Water Pollution Control Federation,* 1969, **41**, (7), 1341–1352.

[8] Heide, B. A. and Pasveer, A. Oxidation ditch: prevention and control of filamentous sludge. *H_2O*, 1974, 7, 373–377.

[9] Tomlinson, E. J. and Chambers, B. *The Effect of Longitudinal Mixing on the Settleability of Activated Sludge.* Technical Report TR 122, Water Research Centre, 1979, 46 pp.

[10] Levenspiel, O. *Chemical Reaction Engineering.* New York and London, John Wiley and Sons, 1962, 501 pp.

[11] Chudoba, J., Ottova, V. and Madera, V. Control of activated-sludge filamentous bulking – I. Effect of the hydraulic regime or degree of mixing in an aeration tank, *Water Research,* 1973, 7, 1163–1182.

[12] Rensink, J. H. New approach to preventing bulking sludge. *Journal of the Water Pollution Control Federation,* 1974, **46**, (8), 1888–1894.

[13] Kroiss, H. and Ruider, E. Comparison of plug-flow and completely-mixed activated-sludge process. *Progress in Water Technology,* 1977, **8**, (6), 169–173.

Discussion

Chairman: **Dr. M. J. D. White** (Water Research Centre, Stevenage Laboratory)

Mr. D. H. Eikelboom (Research Institute for Environmental Hygiene, TNO, The Netherlands) opened the discussion and said that he thought there was one factor missing from Mr. Tomlinson's review. Mr. Tomlinson had talked about activated-sludge but had made no mention of the specific micro-organisms which performed the treatment process. **Mr. Tomlinson** replied that he felt that too much emphasis had been placed upon the identification of specific organisms in the past. He too had once thought that identification was important but he had been disappointed in what he could do with the results. He said there was still a need to determine the effects of operating conditions, plant design and waste water characteristics on the various types of micro-organism. He considered that the easiest way to do this was to look at the effect of changing these parameters upon the sludge settleability.

Miss A. J. Blachford (Water Research Centre) said that Mr. Tomlinson had described the effect of mixing on bulking in a continuous context. She said that quite often bulking occurred for a period, of say six months, but then it disappeared as mysteriously as it had arisen. She asked if he was suggesting that when bulking occurred then one should look for a change in the mixing characteristics in the plant, or should one associate a deterioration in settling characteristics with changes in other parameters.

Mr. Tomlinson said that his final comment in presenting his paper was to ask the conference for suggestions of factors other than the mixing characteristics or deviations from design and operating conditions which could contribute to bulking. He said that there were 'completely-mixed' plants which had no settleability problems. Obviously in these cases there was something that was having a beneficial effect on settleability. He believed that the mixing characteristics of an aeration tank had a greater effect on settleability than any other factor that it was possible to vary, but he would welcome other suggestions from the delegates.

Mr. J. L. Hutton (Grampian Regional Council, Aberdeen, UK) asked if Mr. Tomlinson would like to speculate whether sludge loading rate could be one of the factors which can, under certain conditions, cause filamentous bulking. He cited one small 'completely-mixed' sewage treatment plant, in his region, which had been duplicated. Prior to duplication there had been no experience of filamentous bulking, but as soon as the duplication was implemented and the loading on the plant reduced severe filamentous bulking occurred. **Mr. Tomlinson** replied that he believed there was a relationship between sludge loading and settleability but that there was a different relationship for each plant. He believed it was wrong to assume that having found a particular relationship at one plant changes could be made on another plant with the same results. Mr. Tomlinson referred to Fig. 1 which showed six published correlations between sludge settleability and sludge loading and indicated that there was no universal relationship. He said that were he responsible for operating an activated-sludge plant he would establish the relationship for that particular plant, so that he would know which rates of loading to avoid.

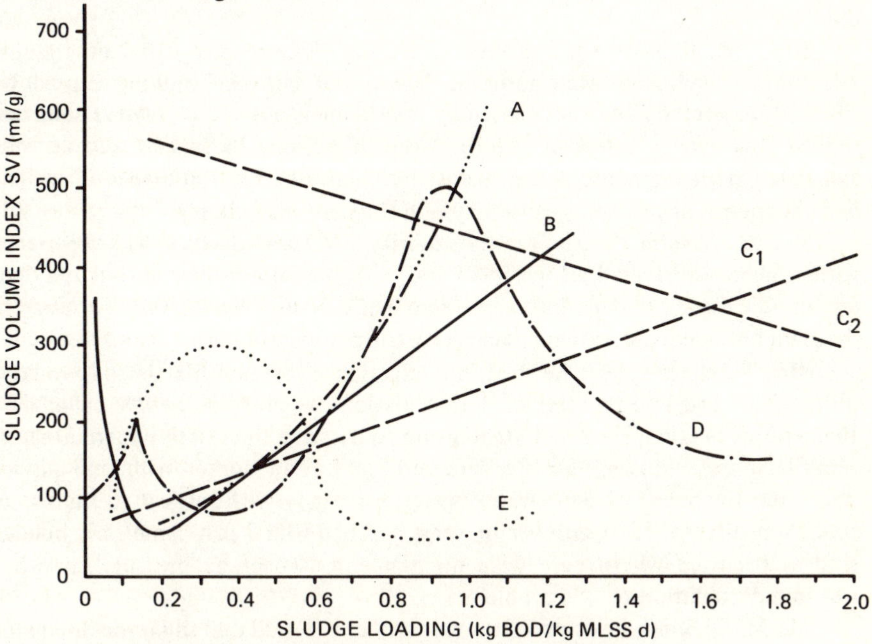

A. Kalbskopf, K.H. *Prog. Wat. Tech. 8* (6) 175–81, 1977

B. Goodman, B.L. *Manual for Activated Sludge Sewage Treatment* Technomic Publishing Co., 1971, 197 pp.

C_1. C_2. Chudoba *et al*. *Water Res. 8*, 231–237, 1974

D. Rideau, J.P., Morfaux, J.N. *La Tech. de l'eau,* (339), (340), 19–35, 17–27, 1975.

E. W.P.R.L. *Water Pollution Res.*, London, HMSO, 1965, 214 pp.

Fig. 1. – Correlations between sludge loading and setteability

Dr. C. F. Forster (University of Birmingham, UK) referred to Fig. 1.1 and said that for an SSVI of about 70 ml/g the Dispersion Number ranged from about 0.007 to 0.20 and if one looked at Dispersion Numbers of 0.10 to 0.20 then the SSVI ranged from 40 to 240 ml/g. He asked if Mr Tomlinson still maintained that it was reasonable to say that SSVI was related to dispersion. **Mr. Tomlinson** replied that he thought it was reasonable to relate Dispersion Number with settleability. He agreed that there was a good deal of scatter on the data, for which there were a number of reasons. One reason was that it was impossible to take any one value of SSVI and say that it was representative of the plant on a long-term basis. Secondly, there was some error involved in the measurement of the Dispersion Number, the technique involved assumed a constant flow rate through the plant whereas the actual flow rate through plants varied during the determinations. If these errors were taken into consideration it was reasonable to say that there was a relationship between SSVI and Dispersion Number. He suggested that if a vertical line was drawn on Fig. 1.1 at a Dispersion Number of 0.10 then most people would prefer to operate plants to the left of the line.

Mr. J. H. Rensink (Agricultural University, Wageningen, The Netherlands) asked Mr. Tomlinson what ratio he had found between bulking caused by filamentous bacteria and that caused by non-filamentous bacteria. **Mr. Tomlinson** replied that most, if not all, of the cases of bulking he had looked at were associated with filamentous organisms. He then asked the audience if anyone had experience of problems caused by non-filamentous bulking.

Mrs. M. Chaplin (Welsh Water Authority, UK) said that she had experience with a plant which had a high SSVI over a fifteen-month period, but only for one month during that period were there many filaments observed. Fortunately the plant had adequate settlement capacity and no operational problems resulted.

Mr. Tomlinson said he had no experience of non-filamentous bulking although he had known cases of deflocculation, which some people called non-filamentous bulking. He asked Mr Rensink to describe the Dutch experience with non-filamentous bulking. **Mr. Rensink** said that he had worked with, and advised upon, the treatment of dairy waste water, especially with respect to the use of oxidation ditches, for a number of years and had found that sometimes bulking sludges occurred when there were no filamentous bacteria present. They had not found a solution to this problem.

Mr. M. G. Booth (Wessex Water Authority, UK) said that so far the discussion had been confined to bulking as defined by Mr. Tomlinson in his paper, but he wondered if Mr. Tomlinson could say whether when he did his survey of settlement problems in 1976 he collected any figures for the relative frequency of deflocculation and rising sludge. He said that his own suspicion was that there were a number of works which had settlement problems which were caused by rising sludge. He asked if Mr. Tomlinson had been able to quantify how frequently that occurred. **Mr. Tomlinson** replied that he did not have the figures that Mr.

Booth requested. He said that the survey was done by use of a questionnaire and that he had to rely upon the answers given. If someone had filled in the questionnaire saying that solids were lost due to bulking, then he had to believe that was the case. He agreed with Mr. Booth that there might be cases where rising sludge or deflocculation were termed bulking. He went on to say that these were specifically excluded from the definition he had given. Some people regarded any form of excessive effluent suspended solids loss as bulking. He believed that there was a need to adopt a more precise definition of bulking.

Mr. B. J. E. Hurley (Thames Water Authority, UK) asked whether, apart from the environmental impact associated with loss of suspended solids, the WRC had been able to identify just how big a problem bulking was and what the cost incurred by the Regional Water Authorities was. He asked to what extent in cash terms did the cost of bulking compare with water quality and health and energy conservation. **Mr. Tomlinson** said he did not have any cost figures but the Water Authorities were continually seeking advice from the WRC on how to overcome bulking problems. His experience was that when solutions were proposed which involved spending money then these solutions were usually rejected and this led him to question how serious the problem really was. If the receiving river was able to cope with the solids then perhaps there was a case for relaxing the discharge standards.

Mr. Hurley then asked about a slide Mr. Tomlinson had shown of data from a survey performed in the USA in the 1940s. He said that the slide seemed to imply that 25% more mechanically-aerated plants, in comparison with diffused-air plants, were susceptible to bulking and asked if this was also Mr. Tomlinson's experience. **Mr. Tomlinson** said that in their most recent survey they had carried out a statistical analysis to determine whether diffused-air plants were any better or any worse than mechanically-aerated plants. The analysis had shown statistically that diffused-air plants produced better settling sludges, but this was not the full story since they had realized that diffused-air aeration tends to be used in 'plug-flow' systems whereas mechanical-aeration tends to be used in 'completely-mixed' systems. They had then performed the work on dispersion that he had described in his paper and arrived at the conclusion that although there was a statistical difference between the plants using the two types of aeration equipment, it was not as a result of any fundamental characteristic of the aeration equipment.

Mr. L. J. Webb (PIRA, Leatherhead, Surrey, UK) said that he had been concerned with activated-sludge plants treating paper-mill effluents where a sludge with good settling characteristics had an SVI of 200 to 300 ml/g, but could go up to 600 to 700 ml/g. He had observed that if the SVI test was continued over a period of say six hours that bulking sludges underwent a period of consolidation of about two to three hours from the start of the test. He said that Mr. Tomlinson had suggested in his introduction that if adequate settlement capacity were provided, solids loss due to bulking could be prevented. He asked if Mr. Tomlinson could comment. **Mr. Tomlinson** said that one of the solutions to the bulking problem was to provide large settlement tanks, but felt that the solutions to be discussed in later papers would probably be less costly than such modifications. He had a feeling, but no evidence, that in some circumstances the longer the settlement time allowed, the worse the settleability became.

Mr. C. B. Waller (Thames Water Authority, UK) commenting upon the problem reported by Mrs. M. Chaplin of non-filamentous bulking, said that some work had been done within Thames Water Authority where a 'plug-flow' system with a reasonable nutrient balance with respect to nitrogen and phosphorus, produced a sludge with a high SSVI value. At times the sludge could have been considered to be bulking. Several possible answers to the problem were examined and they had found that the only unusual characteristic of the sewage was its very low iron concentration. He wondered whether in fact the plant referred to by Mrs. Chaplin has a similar micro-nutrient problem causing poor settleability.

Chapter 2

Study of the causes and prevention of sludge bulking in Germany

F. Wagner*, University of Stuttgart, West Germany

2.1 INTRODUCTION

The experience gained in the causes, prevention and control of bulking has hitherto been almost exclusively confined to dealing with isolated problems. However, achievement of the desired objective is not guaranteed merely by applying control measures successfully adopted elsewhere. In Germany investigations have been made into the waste-water characteristics and operating conditions of more than 100 activated-sludge plants. In addition, research on methods of prevention and control of sludge bulking has been performed. The results have been used in an attempt to account for the variation in settling properties of sludges from the different plants.

2.2 INCIDENCE OF SLUDGE BULKING

Microscopic inspection of about 3500 sludge samples from 315 activated-sludge plants showed that extensive filamentous micro-organism development occurred in at least 45% of them.

In 108 municipal treatment plants sampled during the course of a year, the Sludge Volume Index (SVI) remained at over 150 ml/g for more than six months of the year in 27% of the plants. In more than half the plants the Sludge Volume Index was over 150 ml/g for at least two months of the year. If activated-sludge plants with primary sedimentation and without simultaneous elimination of phosphorus are analysed separately, then at least 70% of them show SVI values greater than 150 ml/g for two months of the year.

The procedures proposed by Eikelboom [1] were used to identify the different filamentous micro-organisms in 356 instances of sludge bulking at 139 activated-sludge plants. Table 2.1 shows the numbers of the individual filamentous micro-organisms causing bulking.

* Now with Wehrle Werk A. G., Energie-und Umwelttechnik, Bismarckstrasse 1–11, Post fach 1110, D–7830 Emmendingen, West Germany.

According to the observations about one in every four bulking phenomena was initiated by the filamentous organism Type 021 N. Either Type 021 N, or *Microthrix parvicella* or Type 0041 takes part in one out of every two phenomena. In spite of its notoriety *Sphaerotilus natans* caused only 9% of the cases of sludge bulking.

Table 2.1 – Occurrence of various filamentous micro-organisms in samples of bulking sludge

Type of filamentous micro-organism	%
Type 021 N	23.3
M. parvicella	15.2
Type 0041	14.6
S. natans	9.0
Nocardia spec.	7.3
H. hydrossis	4.8
N. limicola	4.2
Type 1701	3.4
Type 0961	2.8
Type 0803	2.5
9 other types	6.8
Not definable	6.1

2.3 THE CAUSES OF SLUDGE BULKING

A direct relationship between the occurrence of filamentous activated sludges and the treatment of trade and industrial effluents could be established for the following branches of industry:

- Paper industry
- Milk processing
- Vegetable processing
- Leather and glue industries
- Fruit processing, wine making
- Distilleries.

Effluents from slaughter houses, metal processing and textile industries are less likely to form bulking sludge. Typically, effluents which produce bulking sludges have unbalanced nutrient compostions. In addition, the composition of the sludge has a significant effect on the occurrence of bulking sludge. Filamentous activated sludges are distinguished by their high level of organic compounds, high content of carbon and nitrogen and low content of phosphorus.

As can be seen in Fig. 2.1 there is a clear relationship between the SVI and the ratio of phosphorus to organic Mixed Liquor Suspended Solids (MLSS) compounds.

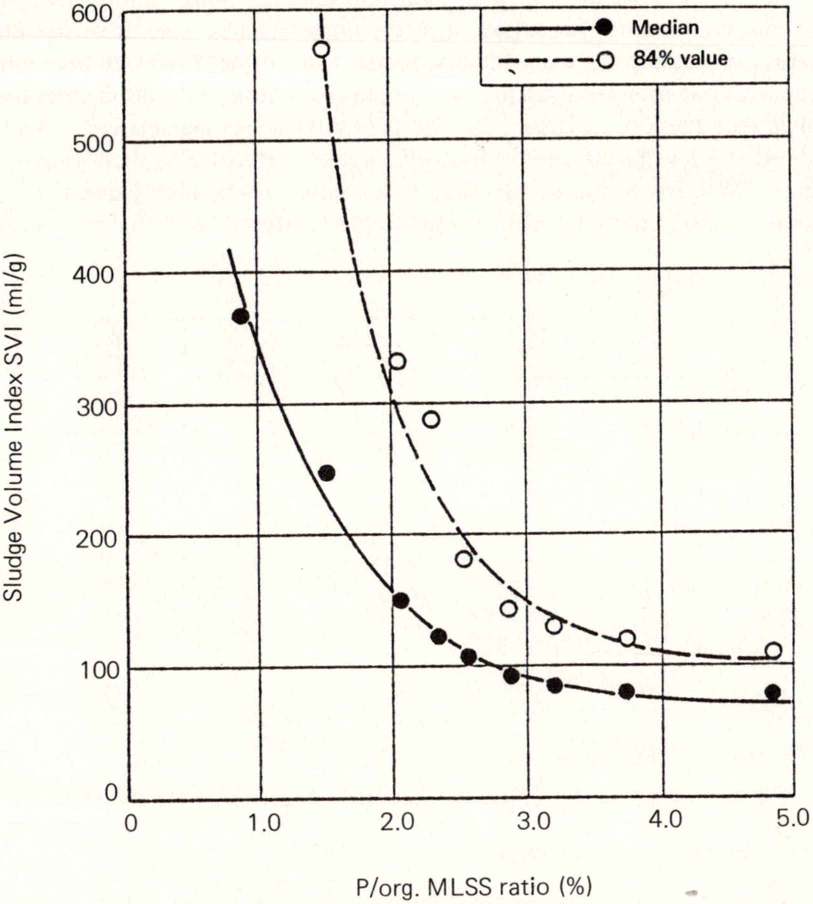

Fig. 2.1 – Relationship between SVI and phosphorus to organic MLSS ratio

A reduction in the phosphorus/organic MLSS ratio to below 3% increases the probability of the formation of highly filamentous activated sludges. However, if this percentage is exceeded bulking rarely takes place.

An explanation of this result might be derived from Pipes' [2] area-to-volume hypothesis. The micro-organisms of the activated sludge satisfy their demand for phosphorus by exploiting the environment surrounding them. Utilization of small concentrations of mainly dissolved phosphate requires development of large bacterial surface areas and hence leads to filamentous growth. The higher the phosphorus concentrations are in the environment of a sludge floc, the smaller is the surface area necessary for adequate assimilation.

Hence, the conclusive step to solve the bulking sludge problem is not to raise the phosphorus concentration in the dissolved substrate, but to fix phosphorus compounds of low solubility to the sludge floc. Compact flocs which settle well have a better capability of fixing and exploiting undissolved phosphorus compounds due to their larger mass and their surface-area characteristics. Accordingly, these flocs profit by selection which gives preference to them in growth. Figure 2.2 shows not only the relationship between settleability and the phosphorus to carbon ratio, but also the importance of the nitrogen to carbon ratio.

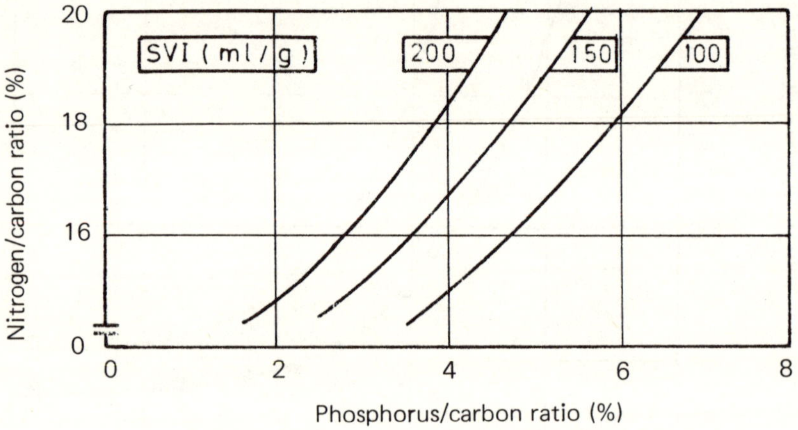

Fig. 2.2 – Relationship between SVI, nitrogen to carbon ratio and phosphorus to carbon ratio

If the nitrogen content of the sludge is low, then a low phosphorus to carbon ratio is sufficient to result in a low SVI. However, if the sludge nitrogen content is high, the phosphorus to carbon ratio must also be high in order to produce a low SVI.

The investigations also revealed a relationship between BOD volumetric loading (kg BOD/m^{3}d) and the SVI. Figure 2.3 shows the maximum SVI occurring between 0.4 and 0.7 kg/m^3 d.

Optimal SVI values occur at BOD volumetric loadings of less than 0.3 kg/m^3 d. Despite several reports in the literature it does not appear that the SVI rises if BOD volumetric loadings are greater than 1.0 kg/m^3 d. It is typical in plants with sludge that settles poorly for nitrification to occur at the critical loading level.

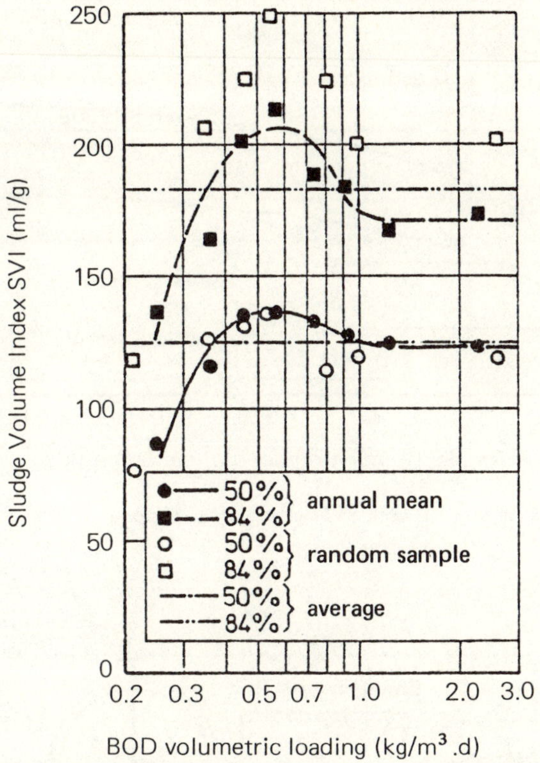

Fig. 2.3 – Relationship between SVI and BOD volumetric loading

2.4 PREVENTION AND CONTROL OF BULKING SLUDGE

Classification of the plants examined into those either with or without primary sedimentation, with simultaneous phosphorus elimination or with pretreatment by biological filtration, yields SVI values as shown in Table 2.2.

Table 2.2 – SVI depending on the mode of operation

Mode of operation	SVI (ml/g)	
	6-months value	10-months value
all plants	103	148
with primary sedimentation	125	183
without primary sedimentation	61	86
with pretreatment by biological filtration	64	92
with simultaneous precipitation	71	87

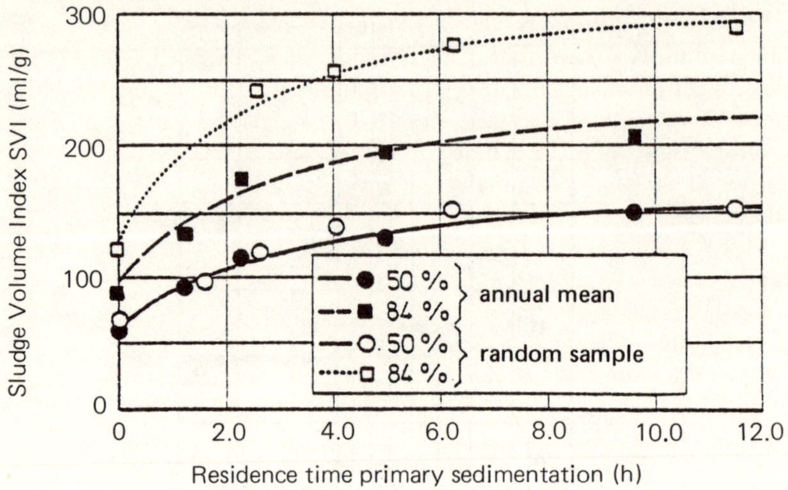

Fig. 2.4 – Relationship between SVI and residence time in the primary sedimentation tank

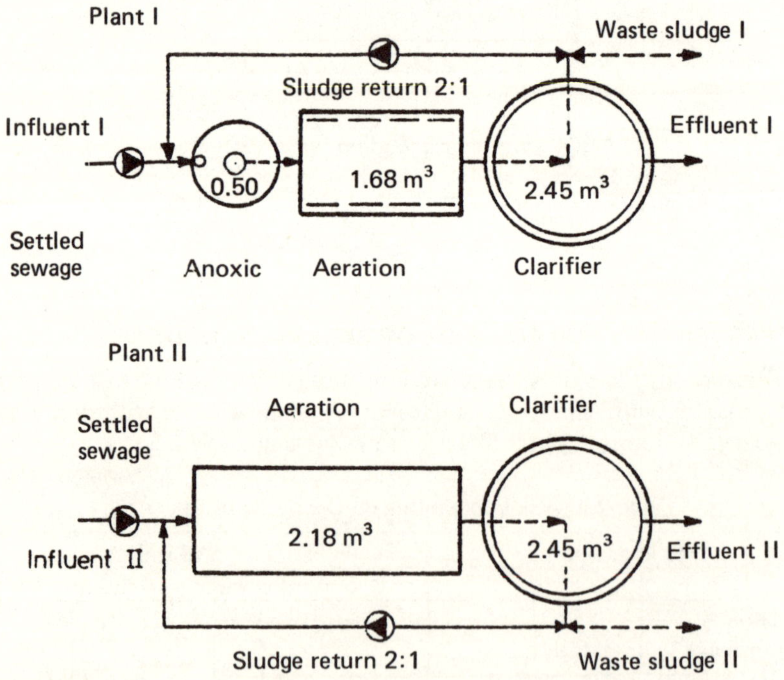

Fig. 2.5 – Flow sheets of the pilot plants with and without denitrification

The worst SVI values by far are obtained in those plants which have primary sedimentation. Plants without primary sedimentation but with pretreatment by biological filtration or simultaneous phosphorus elimination not only have much better SVI values, but they also proved to have considerably less filamentous micro-organisms. Hence the development of bulking sludges is prevented. The influence of the residence time of the sewage in the primary sedimentation tank on the settling characteristics of the activated sludge is shown in Fig. 2.4.

Figure 2.4 shows that the effect on sludge settleability is greatest for primary sedimentation tank residence times of up to three hours.

Further measures for the prevention and control of bulking sludge are:

(i) Installation of a premixing tank in which denitrification occurs.

Two pilot plants (Fig. 2.5) were installed, one of them incorporating denitrification. Both plants were run with a BOD sludge loading of 0.09 kg/kg d.

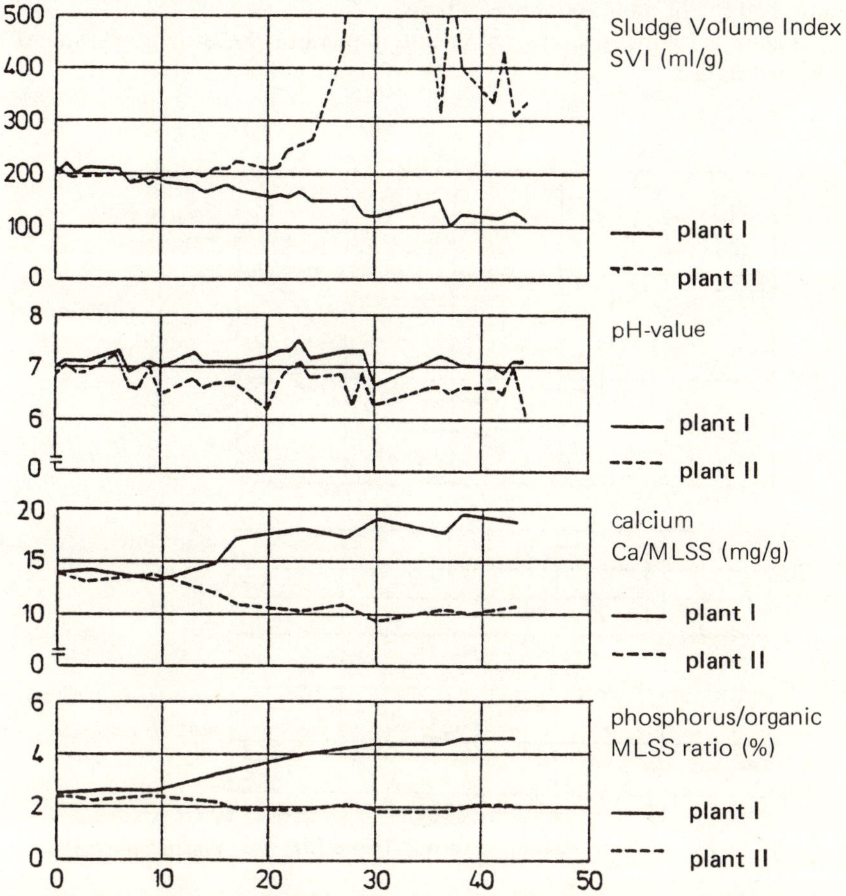

Fig. 2.6 – Variation of sludge characteristics with time. I – with dentrification, II – without denitrification

After 20 days the plant without denitrification produced a very filamentous sludge which was deficient in phosphate. Denitrification in the other plant led to an increase of phosphate in the sludge and relatively good values of SVI. The various phosphorus concentrations in the different activated sludges can be seen to be dependent on the pH-value.

Nitrification causes a reduction in pH-value such that development of calcium phosphate in the sludge is prevented. Contrary to this the release of hydroxide ions due to denitrification supports the production of calcium phosphate in the sludge because of the local increase in the pH-value. This makes an increase in the phosphorus content of the sludge possible.

(ii) Simultaneous precipitation with ferrous salts.
In several municipal sewage treatment plants bulking sludge could be successfully controlled by the addition of precipitants.

Figure 2.7 shows the effect of ferrous sulphate on the settling behaviour of an activated sludge.

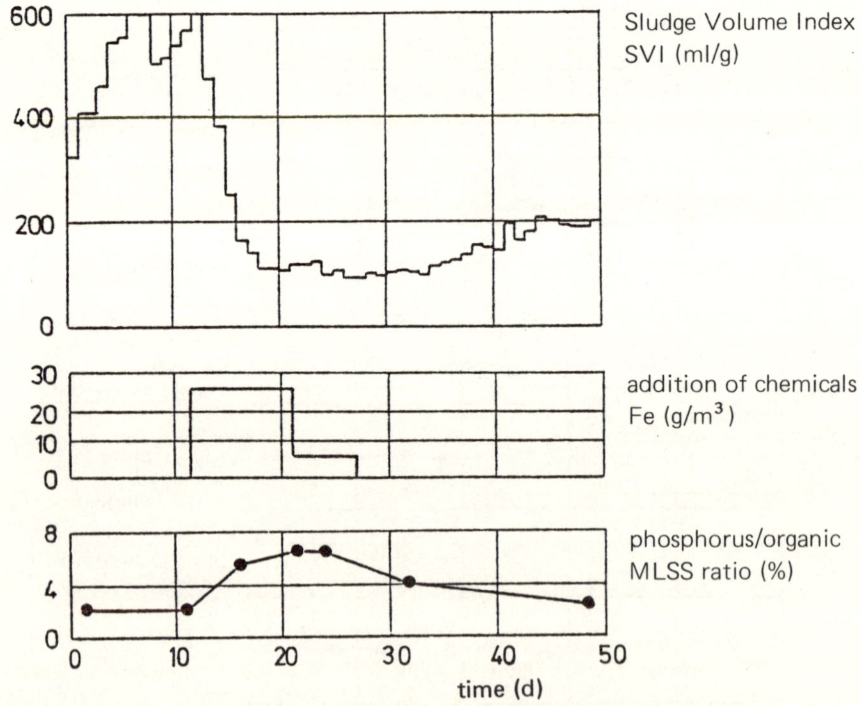

Fig. 2.7 – Variation of sludge characteristics with time during Fe addition

Within a few days filamentous organisms of the Type 0041 could be repressed by the addition of 26 g/m^3 ferrous sulphate. In addition to the coagulant effect, the high sludge phosphorus content which results produces a favourable sludge structure.

In further experiments the addition of ferrous salts also succeeded in affecting other filamentous organisms, namely Types 1701, 0961, 1863 and *Sphaerotilus natans.* However, no reaction was observed for *Microthrix parvicella, H. hydrossis,* Nocardia and Type 0092.

(iii) Simultaneous precipitation with lime.
A significant reduction in SVI was obtained in a municipal sewage treatment plant where calcium hydroxide ($Ca(OH)_2$) was applied (Fig. 2.8).

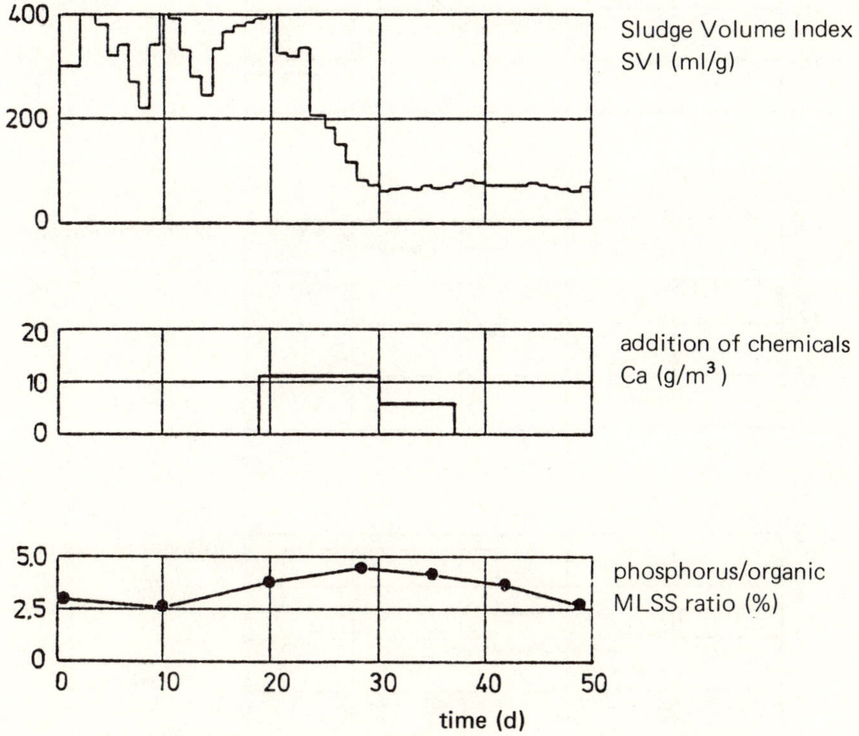

Fig. 2.8 – Variation of sludge characteristics with time during lime precipitation

The filamentous micro-organism Type 021 N could be almost completely repressed within ten days by adding 11 g/m^3 (as Ca) to the influent sewage. The SVI was reduced from more than 300 ml/g to less than 100 ml/g. During the same period phosphorus concentration in the activated sludge increased.

(iv) Addition of basic slag.
The unbalanced nutrient ratio in the effluent from a paper mill near Stuttgart was the cause of bulking sludge in the activated-sludge plant.

The annual average SVI of this plant was greater than 500 ml/g. Using basic slag – a waste product left over from steel production, which is a phosphate fertilizer of low solubility – the settling behaviour of the activated sludge could be considerably improved (Fig. 2.9).

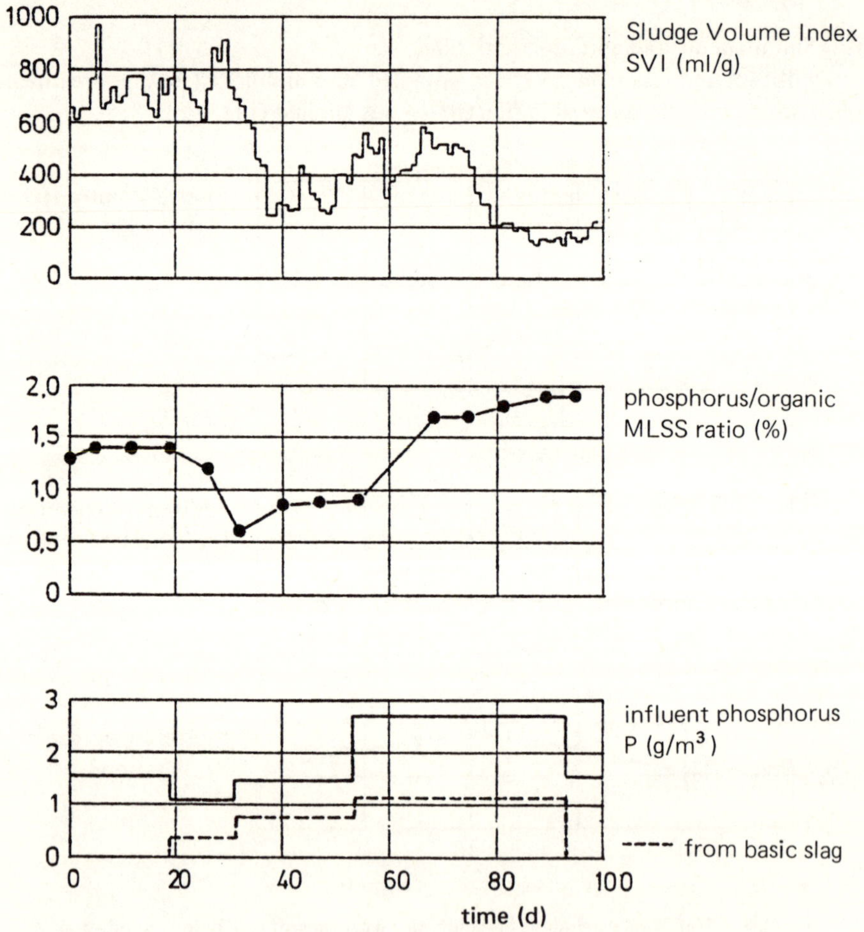

Fig. 2.9 – Variation of sludge characteristics with time during addition of basic slag

The addition of basic slag to the aeration tank influent resulted in an improvement in SVI and this was probably due to a change in the phosphorus content of the sludge. Initially the phosphate supply was restricted and a really significant improvement in SVI was not achieved until the rate of addition of basic slag was doubled. At the same time the major filamentous organism Type 0041 reduced in abundance.

Basic slag is a very dense material which might easily be deposited in channels and tanks, therefore it should be applied only when the hydraulic layout of a plant will allow it.

(v) Application of separators.

The use of automatic centrifugal separators for bulking sludge prevention and control aslo proved successful (Fig. 2.10).

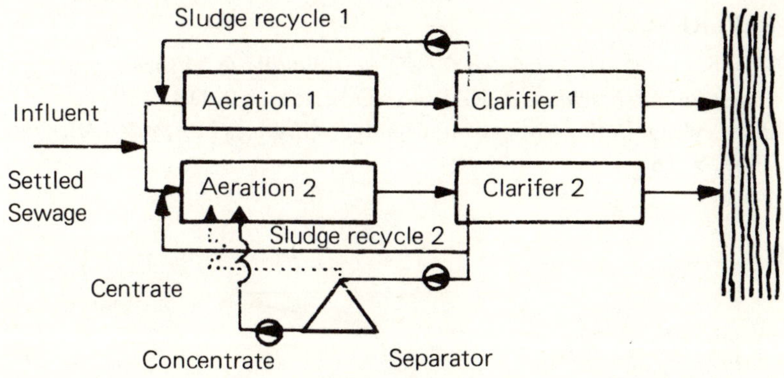

Fig. 2.10 – Flow sheet of the pilot plants with and without centrifugal separation

Feeding the separator with a portion of the sludge recycle flow and discharging the concentrate from the separator back into the aeration tank caused break-up of the filamentous organisms and hence a lowering of the SVI. A deterioration of the plant effluent by loss of sludge could be prevented by this method.

2.5 CONCLUSIONS

The results of the studies in Germany can be summarized as follows:

- Every second activated-sludge plant has problems with bulking sludge.
- The filamentous micro-organisms found most frequently are Type 021 N, *M. parvicella* and Type 0041.
- Activated-sludge plants with primary sedimentation and without simultaneous phosphorus precipitation suffer more bulking-sludge problems than plants without primary sedimentation or with simultaneous phosphorus elimination.
- Bulking sludge is frequently caused by a poor phosphorus supply to the activated sludge.

- Bulking sludge developing in systems with nitrification can be prevented by the provision of denitrification before aeration.
- Bulking sludge can be controlled by adding ferrous salts, lime or low solubility phosphate fertilizer.
- An improved SVI can be obtained by installing separators which cut filamentous micro-organisms mechanically.

The restricted length of this survey did not allow detailed investigation. However, the results show several options for the prevention of activated sludge solids loss into the effluent from final sedimentation tanks.

2.6 REFERENCES

[1] Eikelboom, D. H. *Leitfaden für die Bestimmung fadenförmiger Mikroorganismen im Belebtschlamm.* IGM – TNO Bericht A 9, Delft, 1979.

[2] Pipes, W. O. Bulking of activated sludge, *Advances in Applied Microbiology*, 1967, **9**, 185-239.

Discussion

Chairman: **Dr. M. J. D. White,** (Water Research Centre, Stevenage Laboratory)

Mrs. E. Tramontini (Wessex Water Authority, UK) opened the discussion by saying that she was particularly interested in Mr. Wagner's paper. She had been working at the Palmersford oxygen activated-sludge plant of Wessex Water Authority for the last four years and during this time the effect of various chemicals upon sludge settleability had been investigated. In October 1980 ferric chloride was added to the influent sewage at a concentration of 5 mg Fe^{3+}/1. This was later increased to 10 mg Fe^{3+}/1. The result, as shown in Fig. 1, had been a definite improvement in sludge settleability, which was in agreement with Mr. Wagner's work. In further experiments sodium hydroxide was added to the influent to raise the pH value within the UNOX reactor in an attempt to achieve nitrification. (Carbon dioxide normally accumulates in the enclosed UNOX reactor, leading to depression of the pH value and inhibition of nitrifying bacteria.) Some nitrification was achieved, but it was observed that the sludge settleability improved when the sodium hydroxide dosing system was shut off, as shown in Fig. 2. A relationship between sludge settleability and the percentage of carbon dioxide in the reactor vent gas was established (Fig. 3).

Work performed by Dr. G. P. Green of the Wessex Regional Biology Unit (Chapter 5 of this book) had shown that the addition of sodium hydroxide was associated with a decrease in the filamentous organism Type 0092, but resulted in an increase in number of *Microthrix parvicella* (Fig. 4).

Mrs. Tramontini concluded by asking Mr. Wagner if he had investigated the effect of dissolved carbon dioxide on sludge settleability. **Mr. Wagner** replied that he had not measured carbon dioxide concentration during his work but he had found a good relationship between pH value (and the phosphorus concentration) in the mixed liquor and the control of some types of filamentous micro-organisms. However, these relationships were not reflected by corresponding effects upon the Sludge Volume Index (SVI).

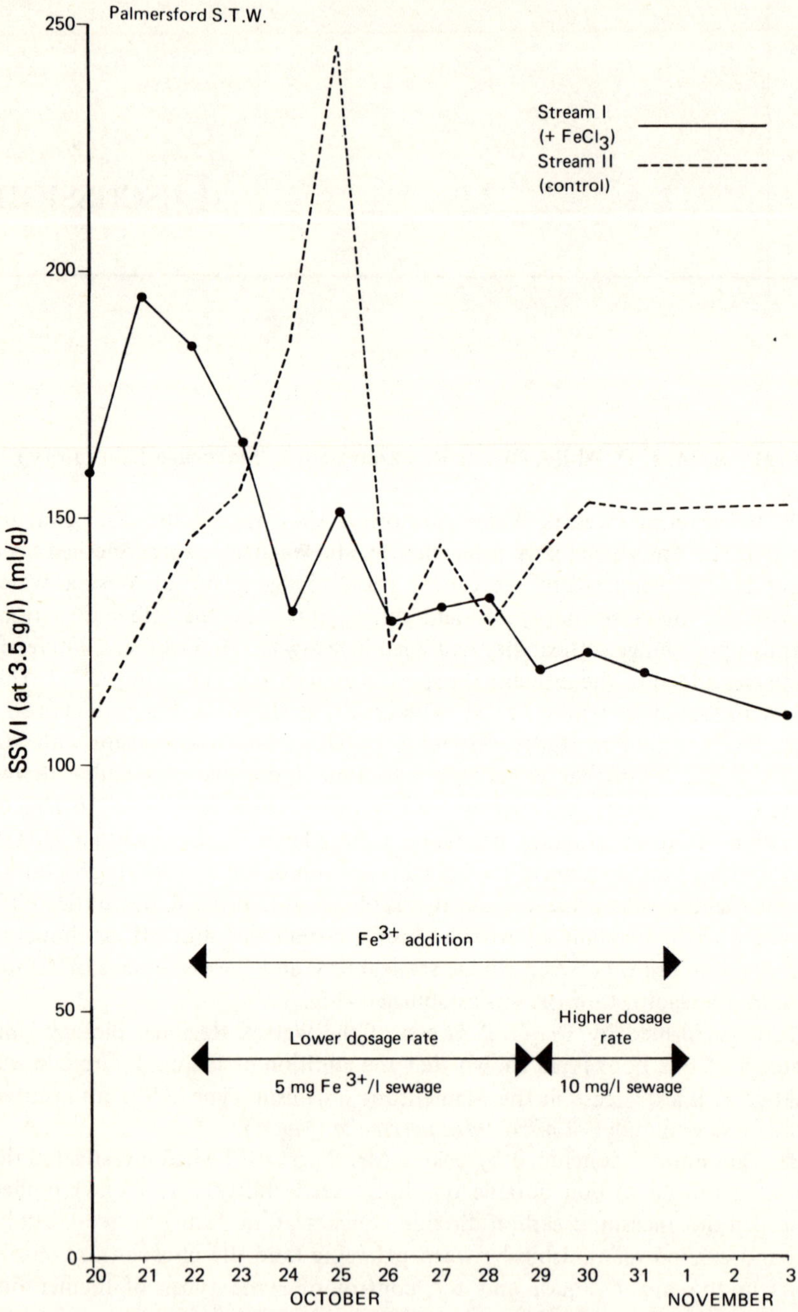

Fig. 1. – The effect of the addition of ferric chloride on sludge settleability – October 1980

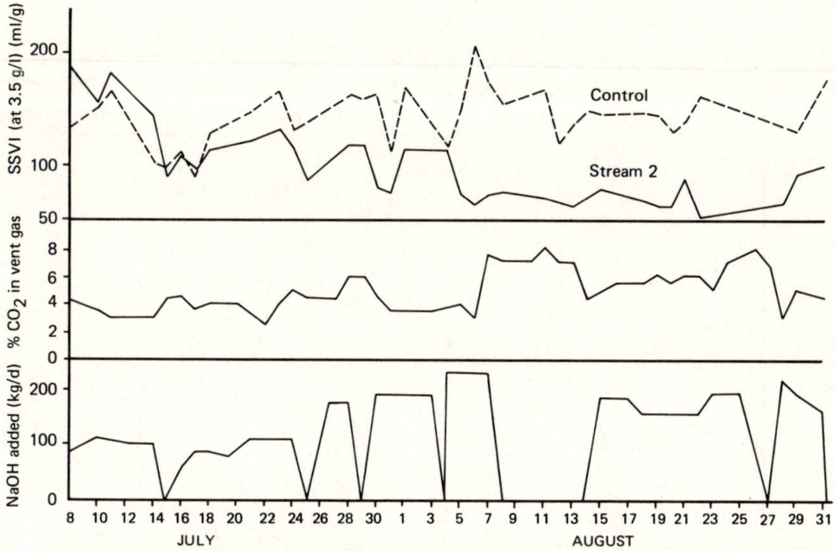

Fig. 2. – Palmersford STW. The effect of the addition of NaOH on CO_2 in vent gas and sludge settleability

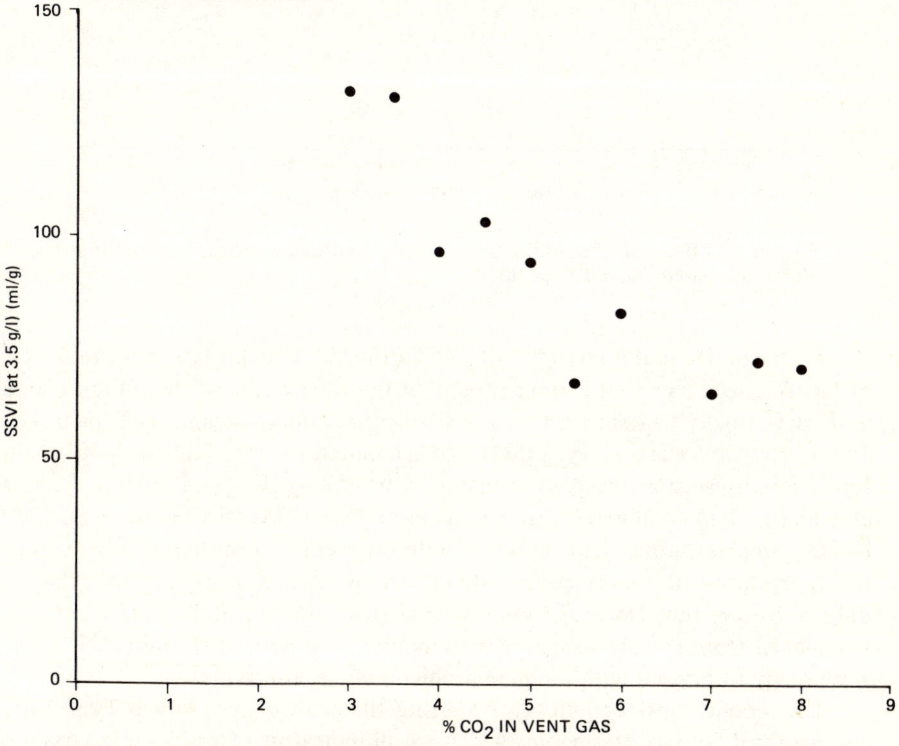

Fig. 3. – Palmersford STW. Settleability (as SSVI) against CO_2 content (%) in vent gas for period 8/7/80 to 31/10/80

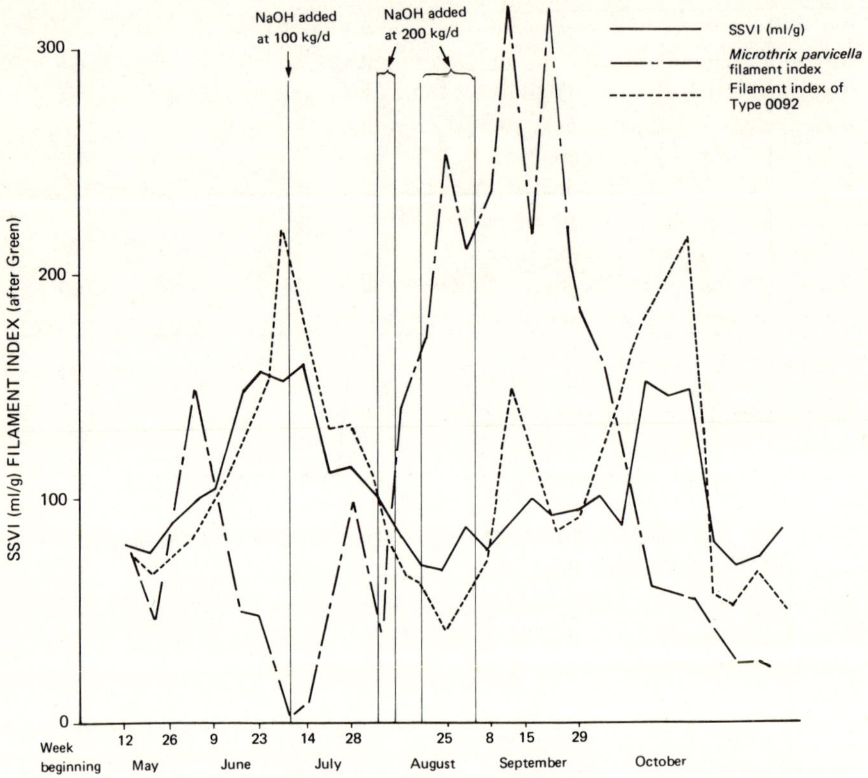

Fig. 4. – Effect of the addition of sodium hydroxide on the populations of *Microthrix parvicella* and Type 0092

Professor D. Jenkins (University of California, USA) referred to Table 2.1 of Mr. Wagner's paper and commented that the survey of activated-sludge plants in West Germany had shown that the filamentous micro-organisms Type 021N, *Microthrix parvicella* and Type 0041 had accounted for over 50% of the filamentous micro-organisms observed in plants with bulking sludges. Professor Jenkins said that he had conducted a similar survey in the USA, on a smaller scale (200 to 300 samples), which had shown a quite different distribution of filamentous micro-organisms, the most commonly occurring being Nocardia. He said that he did not believe that Nocardia was responsible for bulking and would like to see it removed from the list of filaments responsible for poor settleability. Nocardia was usually associated with foaming problems and scum formation.

The second most commonly occurring filamentous species was Type 1701 and the third *Sphaerotilus natans* which are all indicators of low dissolved oxygen (DO) concentrations. He suggested that it might be reasonable, bearing in mind the different organic loadings generally applied in Europe and the USA, that bulking problems in Europe were due to low loadings which might be rectified by some manipulation of the mixing conditions whereas in the USA bulking was usually due to low DO concentrations. He continued by asking Mr. Wagner if he could say, with reference to Fig. 2.1 (which showed an increase in SVI with a decrease of phosphorus concentration in the mixed liquor), what type of micro-organism caused the bulking in that particular case. He also commented that when dealing with bulking of paper-mill wastes it would seem logical to use chlorine or hydrogen peroxide as a method of attacking the problem since most paper mills have an ample supply of these materials on the premises.

Mr. Wagner replied that they had found Nocardia in their survey but not in high frequency; he agreed that it tended to be concentrated in the foam. He said that in the sludge used to produce Fig. 2.1 there were no *Microthrix parvicella* or Nocardia (both of which have high P concentrations) but most of the other types of filamentous micro-organisms were present. Replying to Professor Jenkins' comment about the different possible reasons for bulking in the USA and Europe, Mr. Wagner said he did not know about the situation in the USA but he did know that the situation was similar in Austria to that in Germany. Mr. Wagner said that he had found that when dealing with bulking paper-mill wastes the most frequently occurring micro-organism was Type 0041. He said that in his experience nearly all activated-sludge plants treating paper-mill effluents had bulking sludge and that often they had an unbalanced nutrient ratio in the influent.

Professor P. Grau (Prague Institute of Chemical Technology, Czechoslovakia) commented that he would like to add brewery effluents to Mr. Wagner's list of effluents that promote bulking. Also he thought that the definition of paper-industry effluents should be extended to cover all pulp and paper-industry effluents. He went on to say that he could confirm Mr. Wagner's observation regarding the relationship between phosphorus uptake by the micro-organisms and Sludge Volume Index. He had found the same relationship, namely that the phosphorus uptake was inversely proportional to the SVI.

Mr. J. Watkins (Yorkshire Water Authority, UK) said that his Authority had done some work in which iron salts were added to the activated-sludge process. They had found that the size of the activated-sludge floc did not increase when dosed with iron, but the iron concentration in the sludge could go up ten fold over a seven-day period. He asked if Mr. Wagner would like to comment on this. He then asked if Mr. Wagner could say what effect the addition of phosphorus to an activated-sludge plant which was phosphorus deficient would have upon the respiration rate. **Mr. Wagner** said that, with the exception of systems in which *Microthrix parvicella* was predominant, the addition of iron salts would cause the SVI to decrease. Mr. Wagner said that he had not examined the respiration rate effect of adding phosphorus to a system which was phosphorus deficient, but he expected that the rate would increase.

Professor H. Verachtert (University of Louvain, Belgium) asked Mr. Wagner if he could say how the absence of primary sedimentation could affect bulking. **Mr. Wagner** said that if there were no primary sedimentation then the substrate ratio would be different from that where primary sedimentation was used. He said that there would be more insoluble substrate in the case where primary sedimentation was not used and he thought that this would improve settleability. **Professor Verachtert** then asked if Mr. Wagner meant that the insoluble substrate would add something to the matrix of the sludge floc, perhaps by incorporation of cellulose fibres. **Mr. Wagner** said that he felt that filamentous micro-organisms would not be able to use the insoluble substrate as easily as soluble substrate.

Professor Verachtert then said that Mr. Wagner had stated that small concentrations of dissolved phosphate required the development of large bacterial surface areas. In other words filamentous bacteria with a large surface area established themselves because there was insufficient dissolved phosphate. However, Professor Verachtert said that if the pH value was increased by denitrification or addition of lime, calcium phosphate would be precipitated in the sludge flocs so that even less phosphate would be available (because it was in the insoluble form) and so filamentous bacteria should grow preferentially. This was not the case and so could Mr. Wagner explain this apparent contradiction.

Mr. Wagner said that he believed that during denitrification the rise in pH value, sufficient to cause precipitation of calcium phosphate, was very localized, being limited to the area very close to and inside the floc, and was not as marked in the bulk of the liquid. He also believed that phosphorus in the precipitated form of calcium phosphate would be available to the micro-organisms. **Professor Verachtert** said that he was not sure that the phosphorus in the form of precipitated calcium phosphate would be available to the bacteria. He believed that the precipitated calcium phosphate would increase the density of the floc and improve settleability. He believed this to be the reason for improved settlement, and it was not a matter of eliminating some filamentous bacteria. **Mr. Wagner** said that there were many reports in the literature which indicated that micro-organisms could take up undissolved phosphorus.

Professor Grau said that he believed that the phosphorus which had been precipitated was available to the bacteria because not only was calcium phosphate precipitated but also ferric phosphate, hydroxyapatite and other compounds. He said that it had been shown many times that algae could use phosphate from precipitated salts. He believed that bacteria would also be able to do this.

PART 2: MICROBIOLOGICAL ASPECTS – THE STATE OF THE ART

Chapter 3

Microscopic sludge investigation in relation to treatment plant operation

D. H. Eikelboom, TNO Research Institute for Environmental Hygiene, The Netherlands

3.1 INTRODUCTION

The activated-sludge process is applied on a large scale for the treatment of waste water. Due to the activity of agglomerated bacterial cells – the familiar activated-sludge flocs – a great variety of compounds present in the waste water are decomposed rapidly. The activated-sludge flocs are the hub around which the whole process turns. A poor quality effluent is very often caused directly by sub-optimal conditions in the sludge flocs. Nevertheless, the standard investigative procedure used most often nowadays for checking the course of the treatment process involves chemical and physical analyses of influent and effluent. Thus, plant operators have hardly any information about the actual sludge quality in their plant. Owing to this lack of knowledge they usually find it impossible to interpret a disruption of the process correctly. A turbid effluent, for example, may not only be caused by poor settling of the flocs in the final clarifier (due to an excessive growth of filamentous organisms), but also by too small flocs, by the absence of Protozoa, or a dispersed growth of bacterial cells. Although the final effects of all these failures, viz the turbid effluents, are comparable, there are different causes. It is evident that in the final choice of corrective action to improve plant operation, the real nature of the problem must be known. For this a microscopic examination of the sludge is usually indispensible, because information about the actual quality of the floc may generally be obtained only in this way.

Moreover, regular microscopic examination of sludge is necessary to check the course of the treatment process accurately and allows a deterioration of the sludge quality to be detected before problems occur on a large scale. In this chapter the performance of the examination will be discussed only in brief. A description of the complete procedure is given in a manual on this subject [1].

3.2 MICROSCOPIC EXAMINATION OF ACTIVATED SLUDGE

In an activated-sludge plant the micro-organism population is apparently in equilibrium. However, in reality there are continuous changes. This is not always discernible, on examination of the total treatment process, which is logical since changes in the sludge population will be directed towards making maximum use of the nutrients supplied; in other words the removal of compounds present in the waste water. The relatively stable character of the activated-sludge process is directly related to this. The composition of the floc will, within certain limits continually adjust to changes in the operational procedures, such as quality of the influent and seasonal influences.

It appears from various experiments that the massive growth of many filamentous bacteria is, for example, directly due to the very low nutrient concentration in the aeration tank. An alteration in operating conditions resulting in a locally high loading has therefore been successfully applied for the control and prevention of bulking sludge in many plants. This process adjustment changes the competition between the different sludge bacteria in such a way that the filamentous bacteria can absorb only a small percentage of the nutrients available.

Microscopic sludge examination was originally developed so that changes in the amount and the type of filamentous organisms with time could be recognized and presumptive evidence of filamentous bulking obtained. However, it has been shown that the value of these analyses can be increased considerably, without taking much extra time, when the investigation is not restricted to these organisms. Many other changes in the floc composition can be noticed simultaneously.

To an untrained eye every activated-sludge floc looks like a highly heterogeneous and irregular shaped conglomerate, almost without structure. Here and there particles may be observed moving around, while between the flocs filamentous structures are usually present. However, a more experienced observer will recognize clear differences in relation to:

- shape, structure and size of the flocs,
- composition of the floc: is there a distinct variation of micro-organisms; are characteristic groups of bacteria present; are there many (in)organic non-bacterial particles in the floc?
- filamentous micro-organisms: are these organisms present; what kinds can be distinguished; does the population change with time?
- dispersed growth of bacteria: are many free bacterial cells present between the flocs?
- 'higher' organisms: are Protozoa, rotifers etc. present and how many?
- spirochetes, spirils.

A microscope is a very useful aid when gathering information about all these sludge quality parameters. However, it is no more than that. Microscopic examination gives information on a number of identifiable features of the sludge concerned. Generally little information about the 'biochemical quality' of the floc will be obtained. Thus, the activity of the biomass must be determined by different procedures.

3.3 PERFORMANCE OF THE ANALYSIS AND RECORDING OF THE RESULTS

Nearly every type of activated sludge is so heterogeneous in floc morphology, and floc composition that it is very difficult to describe this complex subject in detail. However, this is not necessary since a rough description gives sufficient information for the characterization of the quality of a sludge. Therefore a gross estimation by eye of most sludge quality parameters is chosen. There is little value in a more exact determination of these parameters by counting and measuring. The total procedure can be performed rapidly; an experienced observer requires no more than about 15 minutes for this.

For the analysis, only a microscope is required. This should be equipped with at least three phase-contrast objectives (magnification: 10, 40 or 50 and 100 times), a 100 x bright field objective for the examination of two bacteriological stains and a calibrated eyepiece micrometer. The greater part of the information required can be obtained directly by close examination of a wet mount preparation of the sludge. Thus, microscopic examination of sludge does not require expensive investments.

The analysis consists of four separate classifications, namely:

(a) filamentous micro-organisms,
(b) Protozoa,
(c) floc morphology,
(d) other characteristics.

A standard method of recording the observations has been developed. Table 3.1 is an example of this. The recorded observations relate to bulking sludge from an oxidation ditch. The upper part of the form is reserved for the description of the population of filamentous micro-organisms; the middle and lower blocks for the Protozoa and floc morphology respectively. The extent to which Protozoa, zoogloea colonies, spirochetes, spirils and organic fibres are present is determined on the basis of a gross estimation by eye of the number of cells or particles in one wet mount preparation. The observations are quantified and recorded as follows:

— = absent
± = occasionally observed
\+ = regularly observed (5 to 15 specimens in one slide)
++ = frequently observed (more than 15 to 20 specimens in one slide).

Table 3.1 – Microscopic examination of a bulking oxidation ditch sludge

FILAMENTOUS MICRO-ORGANISMS(1)				CATEGORY: 4	
"Cyanophyceae"		*S. natans*		Type 0914	
H. hydrossis	0	Thiothrix		Type 0961	
M. parvicella	X	Type 0041	0	Type 1701	
N. limicola I		Type 0092	0	Type 1851	0
N. limicola II		Type 021N			
N. limicola III		Type 0581			
Nocardia		Type 0803		Various species	

PROTOZOA – ROTIFERS – NEMATODES – AMOEBAE(2)					
Cilates	+	*Lionotus*		*Monosiga*	
Carchesium		*Paramecium*		*Pleuromonas*	
Epistylus		*Spirostomum*		*Poteriodendron*	
Opercularia	+	*Trachelophylum*		*Trepomonas*	
Vorticella	+				
Aspidisca				**Amoebae**	–
Blepharisma				**Thecamoebae**	++
Chilodonella		**Flagellates**	–	**Heliozoa**	–
Colpidium		*Bodo*		**Rotifers**	±
Euplotes	±	*Hexamitus*		**Nematodes**	–

MORPHOLOGY OF THE SLUDGE FLOC						VARIOUS FEATURES	
Firmness and shape of the floc	structure: open	structure: compact	size: small	size: medium	size: large	Diversity	+
Firm, somewhat rounded		x	20			Free cells(3)	±
Firm, irregular shape						Zoogloea's(2)	–
Weak, somewhat rounded						Spirochetes(2)	++
Weak, irregular shape						Spirils(2)	–
Agglomerates	x			40	40	Organic fibers(2)	+

For the dispersed growth of bacteria a comparable method is used. For ciliates and flagellates a further subdivision can be applied. Both groups include hundreds of different species and only a specialist in this field can distinguish them from each other. On the analysis form several species are mentioned which frequently occur in activated sludge. It is important to be able to recognize these species because the presence or absence of some Protozoa provides information about the quality of the activated sludge [2,3].

Assessing a precise floc morphology is the most difficult part of the microscopic sludge examination. The flocs are characterized on the basis of:

- dimensions: diameter $< 150\ \mu m$: small
  $150 <$ diameter $< 500\ \mu m$: medium-sized
  " $> 500\ \mu m$: large
- shape : somewhat rounded/irregular
- firmness : firm/weak
- structure : open/compact

After characterizing the different flocs in a sludge in this way, the proportion of the various floc types in the total MLSS content is estimated.

3.3.1 The extent of filamentous growth

Estimation of the sludge parameters related to the bulking sludge phenomenon forms the chief element of microscopic examination of sludge. In almost every activated sludge filamentous micro-organisms (mostly bacteria) are found. They belong to the normal sludge population. As long as their numbers are not too large their presence causes no problems. One might say, on the contrary, as well as contributing to the treatment process they may also give the floc a firm structure. Finally, because of the presence of moderate numbers of filamentous bacteria, the effluent often contains only very small amounts of so-called pinpoint flocs (filtration effect in the final clarifier). Thus, the presence of filamentous bacteria is only a problem when they dominate the sludge population.

The total number of filamentous organisms in a certain sludge can change greatly in the course of time. The rate of these changes depends mainly on the composition of the sewage, the sludge loading and the sludge age. Of course, it is important that these changes can be observed and recorded in some way. But here, a problem arises, since counting and measuring of the filaments is not only very time-consuming, but often it is impossible when the filaments are strongly bent and twisted. Therefore the extent of filamentous growth is also roughly estimated by eye. Activated sludges may be divided into so-called

---

(1) : x = Dominant; 0 = Secondary
(2) : – = absent; ± = incidentally; + = some (5 to 10) cells or particles per preparation; ++ = many cells or particles per preparation.
(3) : – = few; + = some tens per view; ++ = hundreds per view

'categories' on the basis of the extent of filamentous growth. Five categories are distinguished:

0 = filamentous organisms almost absent
1 = small numbers of filamentous organisms
2 = moderate numbers of filamentous organisms
3 = large numbers of filamentous organisms
4 = excessive numbers of filamentous organisms

In ascertaining the category of a certain sludge the microscopic view is compared with a series of reference images in the form of photographs of the different categories (Fig. 3.1). The sludge is given the category number of the photograph which best resembles the microscopic view. In this way, the extent of filament formation is more or less quantified. Application of this system offers many advantages. Subjective descriptions such as 'little', or 'many', which may cause much confusion, are replaced by a number. This is also understood by others, provided they use the same reference images. Moreover, variations occurring in the course of time can be recorded simply using this method.

This method is only meaningful if an increase in the category number actually involves a deterioration in the settling properties of the sludge. Usually, there is such a relationship (Fig. 3.2). A general rule for this cannot be given, however, because the settling properties of activated sludge not only depend on the extent of filamentous growth, but also on the kind of filaments present (large filaments cause more problems than thin, twisted ones), the specific gravity of the floc and the bound water in the floc [4]. A shift of the floc population into filamentous bulking is often observed more clearly, and consequently at an earlier stage, from an increase in the category number than from an increase in the Sludge Volume Index (SVI). If the SVI only is recorded bulking sludge is not identified until the SVI increases significantly. Then the sludge usually contains many filaments already; see, for example, the periods of weeks 32 to 33 and 35 to 39 in Fig. 3.2. Category 3 often forms a kind of critical range. The sludge still settles quite well but a small increase of the number of filaments results in real bulking problems. This rather abrupt increase of SVI as soon as the number of filaments exceeds a given limit has also been noted by others [5].

This method does not seem very exact, but it has proved to be very useful in bulking sludge research. In addition this system of classification can be learned very quickly. A group of people (participants in a three-day course on microscopic sludge examination) were first shown some demonstration samples. Next, they were to determine in about 45 minutes the category number of ten unknown sludges. The results are given in Table 3.2 and are self-evident; they are in very good agreement with each other. This method of standardization is already employed by many water authorities in the Netherlands.

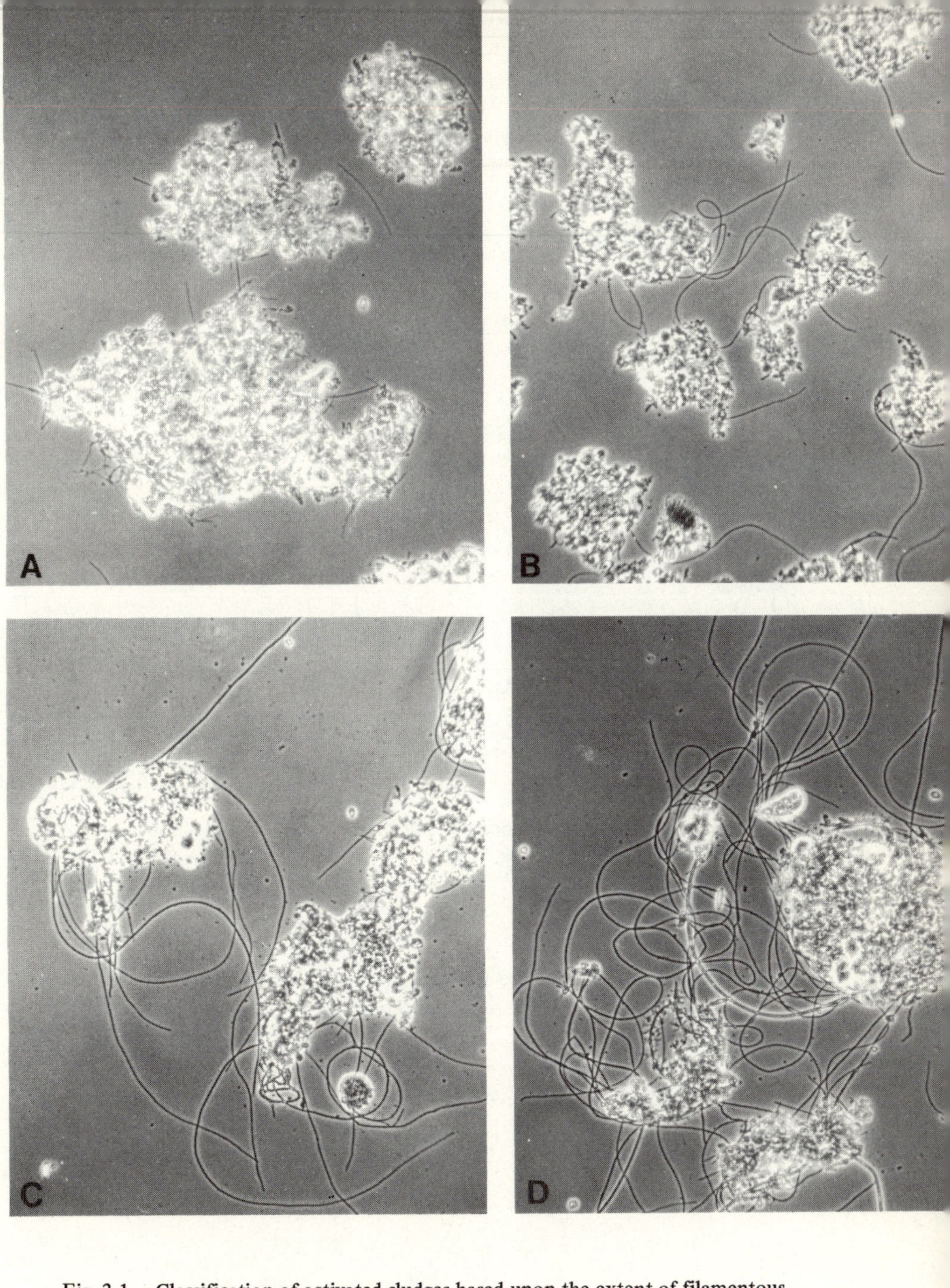

Fig. 3.1 – Classification of activated sludges based upon the extent of filamentous growth. Type 021N dominant; 125 x
A = category 1, B = category 2, C = category 3, D = category 4

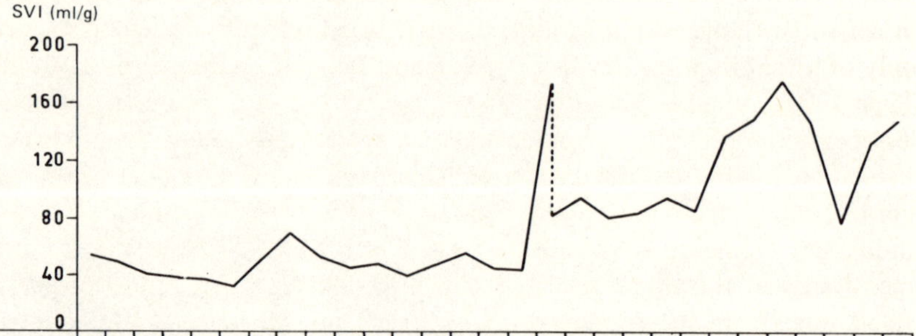

Week nr →	18		20		22		24		26		28		30		32		34		36		38		40		42		44		46
Category	0	0	0	0	0	0	1	3	3	2	0	0	0	0/1	2	2	4/2	2/3	3	2	3	3	3	4	4	4	3	4	4
Filamentous organisms																													
S. natans																		o	x	x	x	x	x	x	x	x	x	x	x
Thiothrix															o	x	x	o	o	o	x	x	o	x	x	o	x	x	x
Type 0041							x	x	x	x				o	x	o	o	o		o	o	o	o	o	o			o	
Type 021N																					o	o	o	o	o	x			o
Type 1701																		o	o	o		o	o	o	o	o	o	o	
M. parvicella																		x	x										
Type 1851															o	o													

Fig. 3.2 – SVI, category numbers and filamentous micro-organisms in an activated sludge. In week no. 34 the pilot plant was restarted with oxidation ditch sludge. X = dominant; 0 = secondary

Table 3.2 – Results obtained for visual estimation of the filament category number by ten people.

Sample no.	Category number estimated									Real value
1	4	3	4	4	4	4	4	4	4	4
2	2	1	1	1	1	0	2	2	2	1
3	2	2	2	2	3	2	3	4	3	2
4	3	2	3	2	2	3	3	3	3	3
5	2	1	1	1	1	1	1	0	2	1
6	4	4	4	4	4	4	4	4	4	4
7	2	2	1	2	2	2	2	2	2	2
8	0	0	0	0	0	0	0	1	0	0
9	4	2	3	3	3	3	3	3	4	3
10	4	4	4	4	4	4	4	4	4	4

3.3.2 Identification of filamentous organisms

Nowadays it is well known that about 30 filamentous micro-organisms may occur in activated sludge [6]. Contrary to what was previously accepted, *S. natans* is only of minor importance. *M. parvicella* and Type 021N frequently cause bulking; Type 1701, Type 0803, *H. hydrossis, S. natans* and *N. limicola III* contribute to a lesser extent to this phenomenon. The other species only rarely develop extensively in plants treating domestic sewage. There is probably no universally applicable control method. At the present time little is known about the sensitivity of the individual filamentous bacteria to a number of bulking sludge control methods mentioned in the literature. It is expected that this gap in knowledge will be filled quickly in the coming years. However, not all filamentous bacteria cause bulking sludge. Certain species (e.g., Type 0914, Flexibacter, Beggiatoa) have been observed in large numbers in sludge which settled excellently. Others, for example Type 0041 and Type 0092 rarely cause bulking in plants treating domestic sewage. Thus, the growth of Type 0041 in the pilot plant (weeks 24 to 26; Fig. 3.2) was not alarming. Thus identification of the filamentous organisms in a specific sludge forms an important facet of the analysis. The various filamentous organisms are primarily distinguished on the basis of a number of morphological characteristics which can be observed in a wet mount preparation, namely:

- branching of the filaments,
- storage of sulphur by the cells,
- motility of the filaments,
- shape of the filaments and the cells,
- dimensions,
- attached growth of unicellular bacteria on the filaments.

In addition, the procedure includes two stains (Gram and Neisser) and a short test to determine whether the bacteria can store sulphur granules in their cells. Then, with the aid of the identification key (Fig. 3.3) the name or the number of the filamentous micro-organisms is determined. Some species which are very seldom observed in sludge have not been included in this key.

In most sludges it is common to observe more than one species. Usually, there are quantitative differences. Species found in large quantities are called dominating. These are marked by an X on the analyses form. The other species (the secondary organisms) are marked with an 0. Figure 3.2 gives an example of the population composition of an activated sludge.

3.4 FUNCTION IN PLANT OPERATION

Although microscopic sludge examination does not take much time, the practical relevance of this kind of investigation is often questioned. Directly coupled to this is the question of how to improve a poor floc quality. How can the results of this analysis be translated into corrective actions in order to upgrade treatment plants? Before considering this in detail, it should be recognized that:

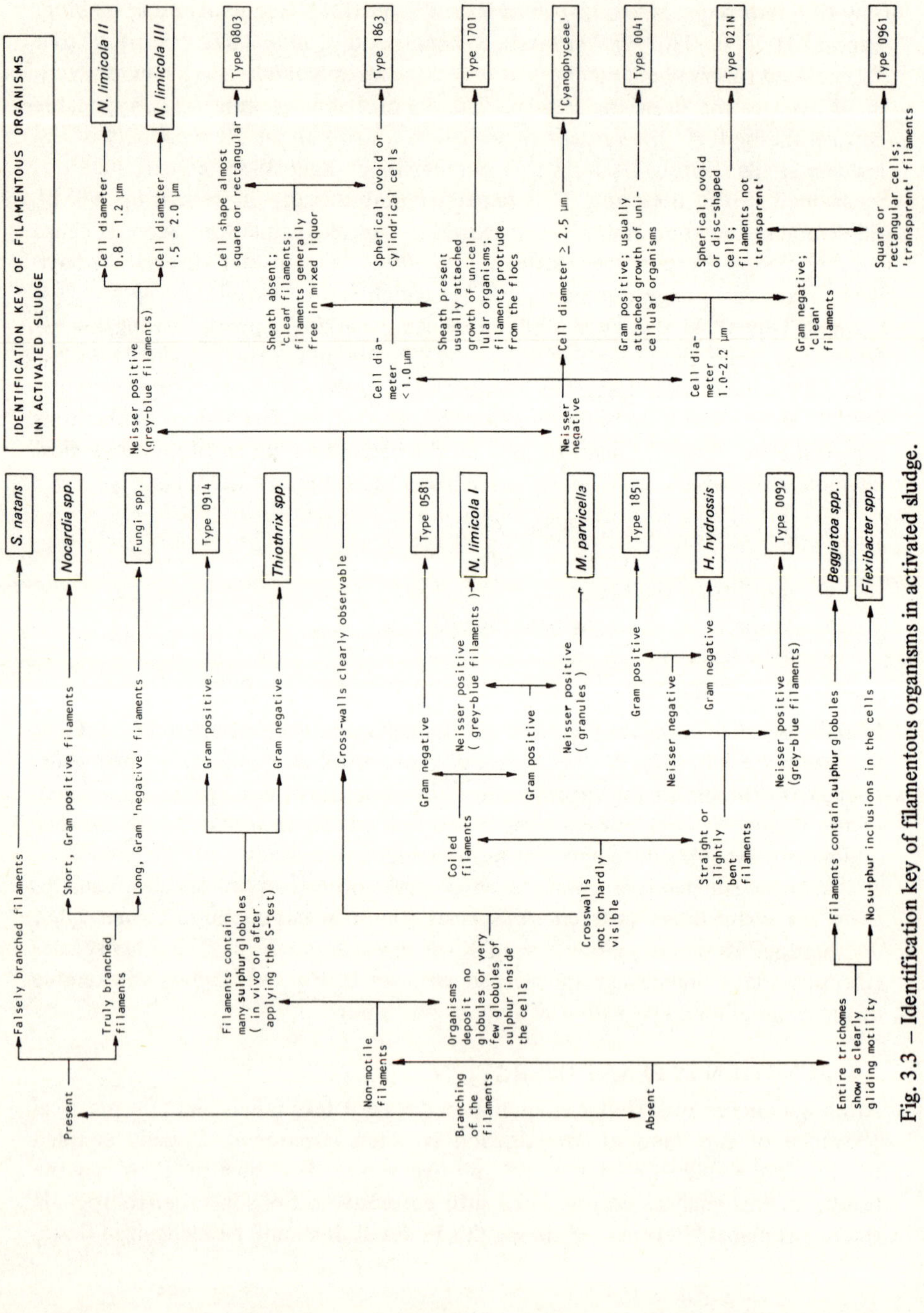

Fig. 3.3 – Identification key of filamentous organisms in activated sludge.

- The bulking sludge problem occurs on a large scale. In the Netherlands, bulking sludge is found in 40 to 50% of all treatment plants treating domestic sewage. In plants treating food-processing waste water this percentage is even higher. In the Federal Republic of Germany and the United Kingdom [7] comparable percentages are found;
- a lot of plants do not meet the design effluent standards. The Environmental Protection Agency has estimated [8] that in the USA about one-third of the plants are not meeting design standards for BOD removal and about half of the plants are not meeting the design standards for suspended solids removal;
- The literature proposes several methods for the improvement of sludge quality.

The results are usually not predictable and in most cases difficult to reproduce elsewhere.

Consequently, there are many unsolved problems. Unfortunately, the knowledge available at the moment is quite insufficient to be able to improve poor plant operation effectively. Very often a trial and error approach is the only available method. Research of a more systematic type needs to be carried out, and more attention should be given to the function of the micro-organisms in sewage treatment. This approach has been recommended several times during the last few years [8, 9, 10, 11, 12] but still operators usually think in terms of a 'tank full of sludge' instead of a 'community of living organisms' [8]. Microscopic examination of sludge is not only necessary to define a problem well, but also to estimate the effect of process changes. In the last few years this systematic approach has been introduced in the Netherlands.

Table 3.3 gives several preliminary results which, at this stage, have not been confirmed definitely by experiments in other plants. The results which have not been published before, originate from various water authorites where this approach is gradually being initiated. In this way the knowledge required to control the growth of filamentous micro-organisms can be built up in a few years.

A similar procedure would have to be applied for solving other operational problems. Although the information on sludge supplied by microscopic examination is still far from complete, steady progress is being made in this field.

Concerning other aspects the following can be remarked:

Protozoa: Generally, it is a good sign when several species of ciliates are found in the sludge. An increase in the number of flagellates and amoebae is usually an indication of a high load and/or insufficient oxygen supply to the aeration tank. Moreover, these organisms also have an obvious indicator function. The protozoan population disappears first if the sewage contains toxic components.

Floc morphology: In low-loaded plants firm, small (50 to 150 μm), compact flocs generally occur; sometimes they form agglomerates. These are built up from small flocs connected to each other by filaments. Sometimes, in very low-loaded plants the flocs disintegrate into particles of 5 to 25 μm diameter which hardly settle. The firmness of the floc decreases as the load increases and irregularly shaped flocs are often observed. Their size, which can vary greatly, depends partly on the aeration system employed.

Table 3.3 – Results of different bulking sludge control strategies

	Dominant filamentous species	Category	SVI (mg/l)	Method applied	Results
A [13]	*S. natans* Type 0961? Type 1851	?	300	Chlorination; 10–20 g Cl_2/m^3 return sludge	Type 0961 and *S. natans* almost disappeared. Type 1851 was not destroyed. SVI dropped to 100.
B	Type 021N	4	400–600	Chlorination; 2.5 g Cl_2/m^3 return sludge	SVI dropped in 14 days to about 125. Type 021N disappeared.
C [14]	*N. limicola III* Type 0041	4	200	Mixing stage of sewage and sludge; floc loading 40–60 mg COD/g MLSS	Category number decreased to 2–3. The SVI dropped to about 100. Dewatering of the sludge improved.
D [15]	*M. parvicella*	4	140–180	Mixing stage of sewage and sludge; floc loading 60 mg COD/g MLSS	*M. parvicella* almost disappeared. Category dropped to 1–2 and the SVI to about 60.
E	Type 021N	4	400–500	Mixing stage of sewage and sludge; floc loading 180 mg COD/g MLSS	Type 021N almost disappeared; category dropped to 1–2.
F	'Cyanophyceae' *H. hydrossis*	4	>300	Chlorination (4g Cl_2/m^3 return sludge) during 1 week followed by the incorporation of a mixing stage; floc loading not known	SVI dropped with Cl_2 to 200. After the incorporation of the mixing stage the SVI decreased to 100–150. *H. hydrossis* disappeared. Moderate numbers of the 'Cyanophyceae' are still present.
G	*M. parvicella*	4	175–300	Mixing stage of sewage and sludge; floc loading not known	SVI dropped to 125–175. *M. parvicella* still present sent (category 3–4)
H	Type 021N *M. parvicella* *N. limicola II*	4	200–400	Mixing stage; floc loading about 120 mg COD/g MLSS	Category 2–3; SVI 100–200. T 021N and *M. parvicella* disappeared.

Diversity: In a 'healthy' sludge, the floc usually consists of many bacterial species. Flocs with a low diversity are often found in very highly-loaded plants and additionally their presence might be an indication of a non-optimum influent composition. A low species diversity makes a plant rather vulnerable because the process depends on the functioning of only a few types of bacteria.

Dispersed bacterial cells: It is normal that in highly-loaded plants the number of free cells between the flocs is considerably larger than in, for example, an extended aeration plant. A rather sudden increase of their number in an activated sludge usually indicates a toxic load or oxygen limitation in the aeration tank. A strong increase of free cells might be succeeded by an explosive growth of Protozoa.

Zoogloea colonies: The massive presence of this species can often be correlated to the composition of the sewage. Their presence usually involves poor settling and dewatering properties of the sludge.

Spirochetes: These bacteria often occur in sludge in multitudes, especially in summer. It is not yet known what significance can be attributed to their presence.

Organic fibres: The presence or absence of these particles is mainly determined by the sewage composition and the functioning of the primary sedimentation tank.

3.5 REFERENCES

[1] Eikelboom, D. H. and van Buÿsen, H. J. J. *Microscopic Sludge Investigation Manual.* IMG-TNO Report A 94A, Delft, The Netherlands, 1981.

[2] Curds, C. R. in *Ecological Aspects of Waste-Water Treatment.* (eds, Curds, C. R. and Hawkes, N. A.) pp. 203–268, Academic Press, London, 1975.

[3] Vedry, B. l'Analyse Ecologique des Boues Activees. *SEGETEC (France),* 1976.

[4] Forster, C. F. and Dallas-Newton, J. Activated sludge settlement – some suppositions and suggestions. *Water Pollution Control,* 1980, **79**, 338–351.

[5] Sezgin, M., Jenkins, D. and Parker, D. S. A unified theory of activated sludge filamentous bulking. *Journal of the Water Pollution Control Federation,* 1978, **50**, 362-381.

[6] Eikelboom, D. H. Filamentous organisms observed in activated sludge. *Water Research,* 1975, **9**, 365–388.

[7] Tomlinson, E. J. *Bulking – A Survey of Activated Sludge Plants.* Technical Report TR 35, Water Research Centre, 1976.

[8] Johnson, W. D. Microbiology: operators should realize its importance in wastewater treatment. *Highlights (Deeds and Data),* 1979, **16**, D13–D14.

[9] Anon. Bulking activated sludge. *Highlights (Deeds and Data),* 1978, **15**, D1-D10.

[10] Pipes, W. O. Bulking of activated sludge. *Advances in Applied Microbiology,* 1967, **9**, 185-234.

[11] Unz, R. F. Micro-organisms contributing to filamentous bulking. *Water Pollution Control Association, Pens.,* 1978, 20-25.

[12] Whittle, T. E. An operators guide to filamentous bulking. *Highlights (Deeds and Data),* 1978, **15**, D1-D8.

[13] Frenzel, H. J. and Sarfert, F. Erfahrungen über die Verhinderung von Blähschlamm Bildung Durch Chlorung des Belebten Schlammes. *GWF – Wasser/Abwasser.* 1971, **112**, 604-606.

[14] Adriaansen, P. L. Practical experience with 2½ year floc-loading in order to correct a poor sludge volume index at the Bunnik Treatment Plant. *De Klaarmeester,* 1978, **5**, 2-6.

[15] Heide, B. A. and Pasveer, A. Oxidation ditch: prevention and contol of filamentous sludge, H_2O, 1974, **18**, 373-377.

Discussion

Chairman: **Dr. E. B. Pike,** (Water Research Centre, Stevenage Laboratory)

Opening the discussion **Mr. C. B. Waller** (Thames Water Authority, UK) expressed interest in the slides shown by the author which showed fibrous material mixed with the activated-sludge flocs. He mentioned an activated-sludge plant within the Thames Water Authority region which treated a paper waste. A number of paper filaments passed into the activated sludge and to his eye looked very much like filamentous bacteria. He asked if Mr. Eikelboom could recommend a method whereby the biological filaments could be distinguished from the paper filaments. **Mr. Eikelboom** described the main distinguishing features, such as cell structure and cell staining, between bacterial and other types of filaments that would be seen through a microscope. However, he recommended that proper training in the identification of filamentous bacteria should be given.

Dr. A. O. Lau (EXXON Research and Engineering, New Jersey, USA) referred to the slides shown by Mr. Eikelboom concerning methods of controlling various types of filamentous micro-organisms. Dr. Lau's research had shown oxygen and dissolved-oxygen concentration to have a significant effect on the growth of *Sphaerotilus natans* and suggested that an additional column should be inserted on the effect of oxygen and dissolved-oxygen concentration. **Professor D. Jenkins** (University of California, Berkeley, USA) reported that he had found that the filamentous bacteria Type 1701 could also be controlled by oxygen. He added that the growth of all filamentous bacteria could be controlled by the addition of chlorine.

Ms. B. Franklin (Yorkshire Water Authority, UK) asked the author how easy it would be for a non-specialist microbiologist or chemist to learn how to identify filamentous bacteria. Secondly, she asked the author to comment on how essential it would be to identify the filamentous bacteria associated with a bulking problem. **Mr. Eikelboom** replied by referring to the text of the paper (Chapter 3 of this book) where it was argued that the microbiology of bulking should receive more attention in order for knowledge of the phenomenon to increase and ultimately for the problem to be solved. He felt there was a need for people on site to learn to distinguish filamentous bacteria so that more results on the effect of different methods of control could be obtained. He added that a non-specialist could learn the techniques in a few weeks.

Professor H. Verachtert (University of Louvain, Belgium) asked the author about two aspects of the identification key for filamentous micro-organisms. He queried the relevance of classification of bacteria on the basis of deposition of no or very few sulphur globules and also on the response to the Neisser reaction as both these could be affected by the physiological state of the bacterial cells. **Mr. Eikelboom** agreed that *Microthrix parvicella* did not always lay down sulphur globules and it was sometimes necessary to add sodium sulphite to the culture in order to obtain a positive response. He agreed also that the Neisser positive granules were sometimes absent in *M. parvicella,* particularly in sludges taken from a highly-loaded plant. **Professor Verachtert** was concerned that the identification key, which was widely used, did not distinguish Type 021N from *Leucothrix* and suspected that some filamentous bacteria reported as Type 021N were in fact *Leucothrix.* In reply **Mr. Eikelboom** agreed that morphologically *Leucothrix* could resemble Type 021N, but he had seen many instances where a bacterium reported as *Leucothrix* was in fact Type 021N. According to his observations *Leucothrix* was rarely present in activated sludge. The final test was to culture the bacteria on specific media.

Mrs. M. Chaplin (Welsh Water Authority, UK) asked the author about the use of gram staining in the identification key for filamentous bacteria. It was known that ageing gram positive cells could exhibit gram negative characteristics. She asked if it should be assumed that these were the same or different bacteria, if both gram positive and gram negative filaments were present in the culture. **Mr. Eikelboom** replied that when a mixture of filaments was seen and the gram negative filaments were similar in all other respects to the gram positive ones, they were assumed to be the same bacterium. **Mrs. Chaplin** continued by referring to another point in the identification key where the presence or absence of a bacterial sheath was a factor. She had found that some sheaths were very thin and difficult to observe under a conventional microscope. She asked the author whether it was intended in the identification key that only very obviously sheathing filaments should be counted as such. In reply **Mr. Eikelboom** agreed that observation of bacterial sheaths could be difficult, particularly when using a phase contrast microscope. The presence or absence of sheaths was an important difference between bacteria. However, the identification key had been constructed in such a way that the presence or absence of a sheath was only of minor importance in distinguishing the filamentous bacteria from each other. He commented that it was generally found that about two to three days' training was required to enable people to use the identification key if they did not have a microbiological background. He mentioned that a course had been set up in Holland for this purpose. After attending such a course, people usually required a few weeks' practise to become experienced in the analysis.

Chapter 4

An attempt to predict bulking in activated sludge by the development of a method to measure filamentous organisms

G. P. Green, Wessex Water Authority

4.1 INTRODUCTION

In September 1979 the Regional Biology Unit of the Wessex Water Authority began a study into the biological aspects of bulking activated sludge. The main aim of the study was to assess the feasibility of using the routine examination of activated sludge, in particular the abundance of filamentous organisms, to predict the onset of bulking so that corrective action can be taken by treatment works' operatives. It soon became apparent that a method of accurately measuring filaments was a necessary prerequisite to further studies. Consequently the main study, which started in January 1980, was divided into two parts. First a programme of sampling activated sludge from the plants operating at six treatment works was carried out to obtain data required for the second and more important part of the study, the development of a method to measure filamentous organisms and its possible use to predict bulking in activated sludge. This chapter describes the work carried out so far which has mainly involved the development of a method to measure filaments and some initial analysis of the resulting data.

4.2 THE SAMPLING PROGRAMME

Samples of activated sludge were taken from the plants operating at the treatment works listed in Table 4.1.

Table 4.1 – List of treatment works sampled

Treatment works	Activated sludge system	Period of sampling
Avonmouth	AB & CD systems	2. 1.80 to 21.10.80
Cheddar		2. 1.80 to 21.10.80
Ham	1 & 2 systems	2. 1.80 to 21.10.80
Holdenhurst	AB, CD & Deep-shaft pilot plant	31.12.79 to 3.12.80
Palmersford	1 & 2 systems	4. 1.80 to 3.12 80
Tarrant Crawford		31.12.79 to 23.10.80

A total of 11 samples from the six treatment works were examined weekly. The samples from Holdenhurst, Palmersford and Tarrant Crawford were processed on the day of sampling, that is within six hours of taking the sample; whilst samples from Avonmouth, Cheddar and Ham were stored overnight at 3 °C and processed the day after sampling, that is within 24 hours of taking the sample. Samples of activated sludge were examined at regular intervals over a 24-hour period and showed no obvious change in character. In addition measurements of temperature, dissolved oxygen and pH were also obtained using a pHOX 62T portable dissolved oxygen and temperature recorder and pHOX 42 portable pH recorder.

4.3 THE ROUTINE EXAMINATION OF ACTIVATED SLUDGE

4.3.1 The routine microscopic examination of activated sludge

A wet preparation of activated sludge was made and examined under a microscope using low power interference contrast (total magnification x 250). The general character of the sludge floc was described and the relative abundance of filamentous organisms assessed subjectively. The activated sludge was then examined using high power interference contrast (total magnification x 625) to identify the different types of filamentous organism present using the key given by Eikelboom [1]. The relative abundance of the different types of filament was subjectively assessed on a scale of 1 to 5. Two staining methods were used as a futher aid to identification of filaments, these were Gram's and Neisser's stains as described by Eikelboom [1].

4.3.2 The routine photomicroscopy of activated sludge

A permanent record of the activated-sludge sample showing the character of the floc and the abundance of filamentous organisms was obtained by taking a series of black and white photographs at lower magnification. Since these would be used in the development of a method to measure filaments, it was necessary to obtain photographs with a high definition but without loosing the true character of the sludge floc and associated filaments. With experimentation the best results were obtained from wet preparations which were on the point of becoming dry. Wet preparations which were allowed to dry completely were unsuitable since the flocs tended to expand and obscure filaments which often contracted within the flocs. Consequently the photographs were a compromise between obtaining the best possible photographic results for measuring filaments, whilst retaining the character of the floc and filaments. Occasionally black and white photographs were taken at high magnification to show the detailed structure of the floc or to record a type of filamentous organism for future reference.

4.3.3 The counts of protozoan ciliates

Counts of the three main groups of protozoan ciliates, the Holotrichia, Peritrichia and Spirotrichia were made using a Sedgewick rafter counting cell. This cell contains 1-ml of liquid and is divided into 1000 squares. Activated sludge was diluted to 1:9 or 1:19 with distilled water depending on the orginal concentration of the sludge. One ml of diluted activated sludge was placed in the counting cell and examined under a microscope using low power phase contrast (total magnification x 125). The number of protozoan ciliates in 60 squares were counted at the 1:9 dilution and in 100 squares at the 1:19 dilution.

4.4 MEASUREMENT OF FILAMENTOUS ORGANISMS

4.4.1 The equipment

After the initial investigation it was obvious that special equipment was required to measure filaments both quickly and with some degree of accuracy. For this a MOP – VIDEOPLAN image analyser was chosen. This is a semi-automatic measuring device with an integrated computer system for the storage of measured values, general calculation and statistical evaluation of measured values, and the display of results. The complete system consists of a basic computer unit, a data monitor, a graphical measuring tablet, a felt-tip marking stylus, and a printer, together with the software to carry out its various functions.

4.4.2 The use of the Videoplan

This equipment can be used in two ways to measure filaments. For the initial evaluation and development of the method the low magnification photographs were used to obtain filament measurements. This is fully described in the next section. The future development of the method for operational purposes will involve using the Video-overlay system. A wet preparation of activated sludge can be viewed on the data monitor by linking the output from a video-camera attached to the microscope and direct measurements of filaments made using the measuring tablet.

4.4.3 The measurement of filaments from photographs

A weekly sample from each activated-sludge plant was recorded by 4 to 6 low magnification black and white photographs. Each photograph was placed on the measuring tablet under an acetate sheet and the filaments measured using the felt-tip marking stylus. Generally no attempt was made to differentiate between the various types of filament except for the patch-forming types *Microthrix parvicella* and Nocardia. In addition the dense areas of floc through which the filaments were not visible were also measured.

4.4.4 The types of filamentous organisms

It was easier to measure certain types of filament than others. The most frequent filaments tentatively identified during the study are listed in Table 4.2.

Table 4.2 – List of filament types identified during the study

Microthrix parvicella	Type 1701
Type 021N	Type 0041
Haliscomenobacter hydrossis	*Sphaerotilus natans*
Type 0092	Nocardia

The thicker and more robust filaments such as Types 021N, 0041 and *S. natans* were fairly easy to measure, whilst greater care was required with finer filaments such as *H. hydrossis, M. parvicella* and Types 0092 and 1701. Greatest difficulty was experienced in measuring *M. parvicella* which usually forms dense patches of tightly coiled thin filaments. Furthermore these patches are often incorporated within the floc. Where *M. parvicella* formed such patches, the areas of those patches that were clearly discernible were measured. Nocardia was measured in similar fashion. In order to express the area measurements of *M. parvicella* in terms of length, a series of high power photographs showing patches of *M. parvicella* were selected. The length of filaments within a total of 25 patches of known area were measured. The result gave a conversion factor of 1 unit area of *M. parvicella* = 0.85 unit length. This figure was used to convert the low magnification measurements of *M. parvicella* from area to length.

4.4.5 The Filament Index (FI)

Since the area of the low magnification photographs over which the filaments could be measured varied according to the area of the dense floc through which filaments were not visible, an attempt was made to standardize the filament length measurements. The total length of filaments from a series of photographs representing a weekly sample was divided into the total area of the photographs over which the filaments could be measured, and then multiplied x 1000 to produce a convenient figure. This Filament Index represents the abundance of filaments outside the floc.

4.4.6 The sludge settleability indices

Data on sludge settleability were obtained from each of the treatment works. Unfortunately it was not always possible to obtain the data as Stirred Specific Volume Index (SSVI) at 3.5 g/l.

4.5 RESULTS

4.5.1 Filament Index and sludge settleability

The relationship between the abundance of filaments (as Filament Index) and sludge settleability (as weekly SSVI (at 3.5 g/l) or SVI) was tested using both the coefficient of correlation (r) and the analysis of variance (F-value). The results are given in Table 4.3.

Table 4.3 – Filament Index (FI): Sludge Settleability (SSVI/SVI)

	Correlation coefficient				Analysis of variance		
Site	r	99%	°F	r^2 (%)	F-value	99%	°F
Avonmouth (AB)	0.49	yes	38	24.5	12.35	yes	38
Avonmouth (CD)	0.29	no	–	8.6	3.57	no	–
Ham	0.17	no	–	2.8	0.88	no	–
Holdenhurst (AB)	0.60	yes	41	35.7	22.81	yes	41
Holdenhurst (CD)	0.87	yes	41	76.1	130.33	yes	41
Holdenhurst (DS)	0.72	yes	42	52.3	46.03	yes	42
Palmersford (1)	0.80	yes	24	63.9	42.58	yes	24
Palmersford (2)	0.11	no	–	1.2	0.29	no	–
Tarrant Crawford	0.45	yes	27	20.3	6.90	yes	27

r = correlation coefficient
r^2 (%) = declared variation as %
99% = significant at 99%
°F = degrees of freedom

Data from Ham (1) and (2) combined as Ham

Data from Cheddar unsuitable for analysis

N.B. at Tarrant Crawford significance tested at 98%

These results are rather inconclusive. There is a good positive relationship between the Filament Index and sludge settleability at Holdenhurst (CD and DS) and Palmersford (1), a moderate positive relationship at Avonmouth (AB), Holdenhurst (AB) and Tarrant Crawford, and apparently no significant relationship at Avonmouth (CD), Ham and Palmersford (2). Generally these results would seem to indicate that there is a relationship between high filament abundance and poor sludge settleability at six of the nine plants investigated.

The relatively poor results from the plants at Avonmouth suggest that factors other than filament abundance may be affecting sludge settleability. Alternatively these poor results may be due to the fact that SVI, the measurement of sludge settleability obtained from Avonmouth, is not always a reliable indicator of sludge settling characteristics.

The poor results at Ham may have been due to operational management techniques involving the addition of ferric sulphate in late May. Although this reduced the abundance of filaments for the remainder of the study period, there only appeared to be a temporary improvement in sludge settleability followed by a gradual rise in SSVI values. The relationship between the Filament Index and sludge settleability was tested for the period prior to the addition of ferric sulphate – see Table 4.4. This produced a very good positive relationship between the Filament Index and sludge settleability, but there was no significant relationship between these parameters for the remainder of the study period after the addition of ferric sulphate. This would suggest that factors other than filament abundance, which were perhaps related to the addition of ferric sulphate, were responsible for poor sludge settleability after late May.

At Palmersford the results from the two plants differed so much that a more detailed analysis of the data was made. At Palmersford (2) this revealed relationships between the Filament Index and sludge settleability previously obscured by considering total filament abundance. At Palmersford it was possible to make separate measurements of the two main types of filament present, the patch-forming *Microthrix parvicella* and the short straight filament *Haliscomenobacter hydrossis.* At Palmersford (2) when these two filamentous species were considered together as a total Filament Index, there was no significant relationship with sludge settleability. However, when the relationship between the Filament Index for each micro-organism and sludge settleability was considered, there was a good positive relationship between the Filament Index for *H. hydrossis* and sludge settleability, but no significant relationship between the Filament Index for *M. parvicella* and sludge settleability – see Table 4.4. This was also the case at Palmersford (1) but here the abundance of *M. parvicella* was so low that it did not obscure the overall relationship between total filament abundance and sludge settleability.

At Tarrant Crawford an improved relationship between the Filament Index and sludge settleability was obtained when the Filament Index for Type 021N was related to sludge settleability – see Table 4.4.

These additional results enabled good positive relationships between filament abundance, either as the Filament Index (FI) or total filament length (FM), and sludge settleability to be established at five plants. Moderate positive relationships were established at three plants. Overall a significant relationship between high filament abundance and poor sludge settleability was established at eight of the nine plants investigated. The results show that measurements of filament abundance are useful as indicators of sludge settleability. The results also suggest that high filament abundance may be a major cause of poor sludge settleability.

Table 4.4 – Filament Abundance (FI/FM): Sludge Settleability (SSVI at 3.5 g/l)

Site	Correlation coefficient				Analysis of variance		
	r	99%	°F	r^2 (%)	F-value	99%	°F
Ham –FI:SSVI Feb 28th–May 22nd	0.87	yes	12	76.2	38.37	yes	12
Palmersford (1) –FM Hh: SSVI	0.82	yes	24	67.4	49.64	yes	24
Palmersford (1) –FM Mp:SSVI	0.23	no	–	5.4	1.38	no	–
Palmersford (2) –FM Hh:SSVI	0.87	yes	24	75.4	73.64	yes	24
Palmersford (2) –FM Mp:SSVI	–0.38	no	–	14.2	3.97	no	–
Tarrant Crawford –FI 021N:SSVI	0.57	yes	27	32.1	12.74	yes	27

r = correlation coefficient
r^2 (%) = declared variation as %
99% = significant at 99%?
°F = degrees of freedom

FI = filament index
FM = total filament length
Mp = *Microthrix parvicella*
Hh = *Haliscomenobacter hydrossis*

Table 4.5 – Total filament length (FM): Sludge Settleability (SSVI/SVI)

Site	Correlation coefficient				Analysis of variance		
	r	99%	°F	r^2 (%)	F-value	99%	°F
Avonmouth (AB)	0.53	yes	38	28.1	14.87	yes	38
Avonmouth (CD)	0.36	no	–	13.0	5.68	no	–
Ham	0.19	no	–	3.7	1.19	no	–
Ham (Feb–May)	0.85	yes	12	71.4	30.01	yes	12
Holdenhurst (AB)	0.62	yes	41	38.9	26.07	yes	41
Holdenhurst (CD)	0.86	yes	41	73.8	115.49	yes	41
Holdenhurst (DS)	0.76	yes	42	57.3	56.31	yes	42
Tarrant Crawford	0.51	yes	27	25.6	9.26	yes	27

r = correlation coefficient
r^2 (%) = declared variation as %
99% = significant at 99%?
°F = degrees of freedom

N.B. data from Palmersford not available

4.5.2 Total filament length and sludge settleability

The results in Table 4.3 and part of Table 4.4 involved using the Filament Index, the calculation of which requires measurements of the zoogloeal floc area, (see Section 4.4.5). For simplicity the measurements of total filament length (FM) were used to test the relationship between filament abundance and sludge settleability. The results are given in Table 4.5.

When these results are compared to those in Table 4.3 and for Ham (February to May) in Table 4.4, there is little difference. This shows that for practical purposes the Filament Index (FI) as described in Section 4.4.5 is not necessary, and that the straightforward measurements of total filament length (FM) will suffice.

4.5.3 Using filament abundance to predict sludge settleability

To assess the use of measurements of filament abundance to predict the onset of poor sludge settleability, the relationship between filament abundance (as FI or FM) and the *following* week's sludge settleability was tested. The results are given in Table 4.6.

Table 4.6 – Filament Abundance (FI/FM): Following Weeks Sludge Settleability (SSVI/SVI)

	Correlation coefficient				Analysis of variance		
Site	r	99%	°F	r^2 (%)	F-value	99%	°F
Avonmouth (AB)	0.29	no	–	8.2	3.39	no	–
Avonmouth (CD)	0.07	no	–	0.5	0.18	no	–
Ham	0.23	no	–	5.2	1.53	no	–
Holdenhurst (AB)	0.43	yes	40	18.5	9.07	yes	40
Holdenhurst (CD)	0.70	yes	39	49.1	37.59	yes	39
Holdenhurst (DS)	0.77	yes	41	59.8	60.89	yes	41
Palmersford (1)	0.71	yes	24	51.0	25.00	yes	24
Palmersford (2)	0.09	no	–	0.9	0.21	no	–
Tarrant Crawford	0.49	yes	25	24.3	8.01	yes	25
Palmersford (1) –FM Hh:SSVI	0.68	yes	24	46.5	20.82	yes	24
Palmersford (2) –FM Hh:SSVI	0.59	yes	24	35.3	13.10	yes	24
Tarrant Crawford –FI 021N:SSVI	0.57	yes	25	32.4	11.99	yes	25

r = correlation coefficient
r^2 (%) = declared variation as %
99% = significant at 99%?
°F = degrees of freedom

FI = filament index
FM = total filament length
Hh = *Haliscomenobacter hydrossis*

These results show that for all plants except Avonmouth (AB and CD) and Ham there was still a positive relationship between filament abundance and the *following* week's sludge settleability. This suggests that measurements of filament abundance have potential for predicting poor sludge settleability.

4.5.4 Counts of protozoan ciliates and sludge settleability

The relationship between the counts of protozoan ciliates and both sludge settleability (as SSVI or SVI) and filament abundance (as FI and FM) were tested using both the coefficient of correlation (r) and the analysis of variance (F-value). The results from all the plants show no significant relationships. Thus total ciliate numbers are not useful indicators of sludge settleability.

4.6 DISCUSSION OF RESULTS

Since activated-sludge plants involve biological processes, it seems logical to suggest that any problems experienced are of a biological nature. High numbers of filamentous bacteria have frequently been considered to be the cause of poor sludge settleability and bulking. The results of this study confirm the association between high filament abundance and poor sludge settleability at eight of the nine plants investigated. This suggests that high filament abundance is a major cause of poor sludge settleability.

Since the various types of filament have different physical characteristics, it seems likely, assuming that filaments do cause poor sludge settleability, that the effect of the different filament types on sludge settleability will vary. There is evidence to support this from the data obtained at Palmersford where the presence of *H. hydrossis* was closely associated with poor sludge settleability, whilst *M. parvicella,* even at high levels of abundance at Palmersford (2), had no apparent effect. However, it should be pointed out that *M. parvicella* was closely associated with poor sludge settleability at Holdenhurst (CD and deep-shaft pilot plant).

The results also indicate that there may be other factors causing poor sludge settleability. Observations on the behaviour of the zoogloeal floc in a Sedgewick rafter counting cell suggest that these factors may involve the physical characteristics of the individual flocs, such as size, shape and density. The presence of very small individual flocs, often referred to as 'pin-floc' has been considered to be a cause of poor sludge settleability. In the settlement of activated sludge there may be a close relationship between the abundance of filaments and the physical characteristics of the zoogloeal floc.

4.7 USE OF THE FILAMENT INDEX

If the Filament Index (FI) in its final form proves to be a reliable indicator of sludge settleability, its main attributes would be as follows:

(i) where filamentous bacteria are associated with activated sludge the FI will predict or indicate the potential for poor sludge settleability;
(ii) in the event of poor sludge settleability the FI will show whether this is caused by high filament abundance or not;
(iii) if poor sludge settleability is due to high filament abundance, it will be possible to identify and perhaps measure the adundance of the different types of filament involved;
(iv) where remedial action is taken, either by changes in the operation of the plant, or by adding chemicals, the effects on the filaments can be measured and carefully monitored.

4.8 FUTURE WORK AND RECOMMENDATIONS

The primary aim of any future work should be to develop the Filament Index (FI) as described above to produce a quick, simple, reliable and cost-effective method of monitoring and predicting poor sludge settleability for operational purposes. Such work should involve:
(i) Developing the use of the video-overlay mode of the Videoplan which enables direct measurements of filament abundance to be made.
(ii) Assessing the minimum number of measured samples required to obtain a reliable result.
(iii) Increasing the frequency of sampling to improve the method for predictive purposes.
(iv) Investigating whether measurements of the size, shape and density of the individual zoogloeal flocs may also be useful as indicators of sludge settleability separately or integrated with the filament index (FI).

4.9 ACKNOWLEDGEMENTS

The author wishes to express his gratitude to the Wessex Water Authority for the facilities they provided to enable this research to be done. The opinions expressed in this chapter are those of the author and do not necessarily represent those of the Wessex Water Authority.

4.10 REFERENCES

[1] Eikelboom, D. H. Filamentous organisms observed in activated sludge *Water Research,* 1975, **9**, 365–388.

Discussion

Chairman: **Dr. E. B. Pike** (Water Research Centre, Stevenage Laboratory)

Dr. C. F. Forster (University of Birmingham, UK) opened the discussion by referring to the statistical methods used in Dr Green's work. He asked how the coefficient of correlation was calculated and suggested that it might not be the appropriate statistic and that coefficient of determination should have been used to relate the sets of data. He had calculated the coefficient of determination on the data presented in the paper (Chapter 4 of this book), omitting results from Holdenhurst AB lanes and the nitrification lane at Palmersford Sewage Treatment Works (given as Palmersford (2)), and found that the percentage of variation in settleability that could be accounted for by the Filament Index was less than 50 per cent. In conclusion he felt that the results presented did not show statistically that Filament Index could be correlated with settleability of activated sludges. Dr. Forster continued by commenting on the identification by the author of the filamentous bacterium Type 0092 as the micro-organism causing deterioration in sludge settleability at Palmersford Sewage Treatment Works, rather than other filamentous bacteria present, in particular *Microthrix parvicella.* He referred to a previous speaker who had said that Type 0092 was not a major cause of activated-sludge bulking. In reply **Dr. Green** agreed that further testing and statistical analysis of data were required, however, he felt that the results did indicate that filamentous micro-organisms were an important cause of poor sludge settleability at these works.

Referring to the method adopted by the author to calculate the abundance of filamentous bacteria, **Dr. Forster** said that Mr. Eikelboom had used a purely visual technique for estimating filamentous growth. He mentioned work on poor sludge settleability at Saltford Sewage Treatment Works where Mr Dallas-Newton (Wessex Water Authority) and himself had also used a purely visual estimation of abundance of filamentous growth with some success. He asked for Dr. Green's comments on the technique described in his paper using computers compared with previous techniques. **Dr. Green** said that the technique using a computer would need further testing and development before it was suggested as a general technique. **Dr. E. B. Pike** (Water Research Centre) commented that the correlation of data presented by the author seemed to be typical of a biological system which generally had many variates.

Describing one of the sewage treatment works on which Dr. Green had measured Filament Index, **Mr. R. A. Varley** (Wessex Water Authority, UK) said that the A and B aeration lanes at Holdenhurst Sewage Treatment Works, which were operated at a high sludge loading, with the surface aeration system supplemented by injection of oxygen, were predominantly used for carbonaceous oxidation. The C and D aeration lanes and also the deep shaft pilot plant on the site achieved virtually complete nitrification. He reported that it had been found that the deep shaft pilot plant and the C and D aeration lanes were very similar in their microbiological content. He asked Dr. Green to comment on the change in microbiology between the two plants that had been seen when an anoxic zone had been introduced prior to the deep shaft pilot plant. In reply **Dr. Green** said that the effect of an introduction of an anoxic zone prior to treatment in the deep shaft pilot plant was a marked decrease in the abundance of *Microthrix parvicella* which occurred at the same time as the improvement in settleability of the activated sludge.

Chapter 5

The physiology of Microthrix parvicella, a filamentous bacterium isolated from activated sludge

H. Slijkhuis and **M. H. Deinema**, Agricultural University, The Netherlands

5.1 INTRODUCTION

Filamentous bacteria, if present in large amounts in activated-sludge flocs, may be responsible for the bulking of the sludge. Van Veen [1] and Eikelboom [2] have studied the dominant filamentous bacteria found in various sludges. Depending on the presence or absence of a sheath, these bacteria can be divided into two groups. The physiological properties of *Sphaerotilus natans* and *Haliscomenobacter hydrossis,* two sheathed bacteria, have been studied extensively in the last ten years [3-8]. However, the main characteristics of the other filamentous bacteria found in bulking sludge are still unknown.

The growth of a micro-organism in activated sludge is determined by the chemical and physical nature of the influent and the type of treatment process. The abundant growth of filamentous bacteria in activated-sludge plants should be correlated with their physiological properties, but even with *Sphaerotilus natans* and *Haliscomenobacter* spp. this procedure is still difficult.

In The Netherlands, bulking of sludge in low-loaded activated-sludge plants and oxidation ditches is mainly caused by *Microthrix parvicella.* This non-sheathed bacterium has been isolated by Van Veen [1] and by Eikelboom [2]. In 1969 Pasveer [9] described a case of bulking due to abundant growth of *M. parvicella,* which he erroneously identified as a filamentous form of *E. coli.* Several strains of *M. parvicella* have been isolated in this laboratory and cultivated on sludge hydrolysate medium [2]. Cell yield was low and growth was not homogeneous due to floc formation. The sludge hydrolysate contained colloidal particles which interfered with the cell yield.

In this chapter a new medium for cultivating *M. parvicella* isolated from an oxidation ditch is described and a number of physiological characteristics of the organism are given.

5.2 RESULTS AND DISCUSSION

5.2.1 The culture medium for *Microthrix parvicella*

Many chemically defined and complex media have been tested which resulted in the medium as given in Table 5.1.

Table 5.1 – Composition of the medium for cultivating *Microthrix parvicella*

Tween 80 (polyoxyethylenesorbitan mono-oleate; Sigma)	4.00 g/l
Bacteriological peptone (Oxoid)	4.00 g/l
$(NH_4)_2SO_4$	0.80 g/l
K_2HPO_4 (1)	8.20 g/l
KH_2PO_4 (1)	0.35 g/l
Solution I (trace elements and $MgSO_4$)(2)	10.00 ml/l
Solution II (trace elements and $CaCl_2$)(3)	10.00 ml/l
Vitamin solution(4)	5.00 ml/l

(1) Potassium phosphate is sterilized separately.
(2) Solution I contains (g/l): $MgSO_4 . 7H_2O$, 7.5; $MnSO_4 . H_2O$, 0.3; $ZnSO_4 . 7H_2O$, 0.01; $CuSO_4 . 5H_2O$, 0.01.
(3) Solution II contains (g/l): $CaCl_2$, 5.0; $FeCl_3 . 6H_2O$, 0.5; H_3BO_3, 0.01; $Na_2MoO_4 . 2H_2O$, 0.005; $CoCl_2 . 6H_2O$, 0.005.
(4) Vitamin solution contains (g/l): Ca pantothenate, 10^{-2}; niacin, 10^{-2}; cyanocobalamin, 5.10^{-4}; folic acid, 5.10^{-4}; pyridoxine, 10^{-2}; *p*-amino benzoic acid, 10^{-2}; inositol 10^{-2}; thiamine, 10^{-2}; riboflavin, 10^{-2}. This solution is sterilized by Seitz filtration and added aseptically to the cooled medium.

Oleic acid (0.05 g/l) proved to be an essential carbon compound; it could be replaced by Tween 80 (polyoxyethylenesorbitan mono-oleate), which is soluble in water. Other organic substrates were the polyoxyethylenesorbitan derivates of stearic and palmitic acids, Tween 60 and Tween 40, respectively. Glucose, fructose, lactic acid, succinic acid and citric acid, which are in general excellent substrates for micro-organisms are not utilized by *M. parvicella.* Ammonium nitrogen is the main N-source, as was shown in an experiment with $(N^{15}H_4)_2SO_4$ in which at least 75% of the labelled nitrogen was incorporated into cell material. Peptone (4 g/l), which can be replaced by yeast extract but not by Casamino acids, was hardly used as a C- or N-source. With smaller amounts of peptone the cell yield decreased suggesting that this nutrient contains an essential but still unknown growth factor.

The nutrient solution was buffered at pH = 8.0 by 50 mM potassium phosphate. *M. parvicella* acidifies the medium to pH = 7.4 during growth. Table 5.2 shows an unusual pH dependency; the sharp drop of cell yield to zero at pH = 7.1 is probably due to the inhibitory effect of liberated undissociated oleic acid. In a medium with oleic acid (initial concentration 0.075 g sodium oleate/1) the sharp drop even occurred at pH = 7.6. Respiration of oleic acid by washed cell suspensions is also highly pH-dependent (Section 5.2.3).

The requirement for the supplied trace elements and vitamins has not yet been studied.

Table 5.2 – Effect of pH on growth of *Microthrix parvicella* in Tween 80/peptone medium

Culture	1	2	3	4	5	6	7
Initial pH	5.7	6.4	6.75	7.1	7.5	7.7	8.0
Final pH(1)	5.7	6.4	6.75	7.1	7.1	7.4	7.6
Cell yield, dry weight, g/l(1)	0	0	0	0	1.25	1.68	1.65

(1) After 6 days at 25 °C

5.2.2 Growth of *Microthrix parvicella*

The Total Organic Carbon (TOC) content of the medium changes slightly during growth (Fig. 5.1); the polyoxyethylenesorbitan part of the Tween 80 molecule and the main part of the peptone remain unchanged. These compounds accounted for 84% of the initial TOC content.

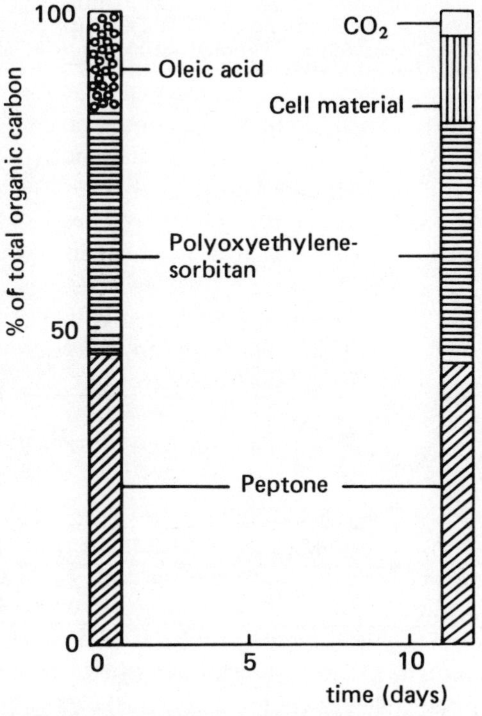

Fig. 5.1 – Distribution of total organic carbon in a culture of *Microthrix parvicella* grown in Tween 80/peptone medium

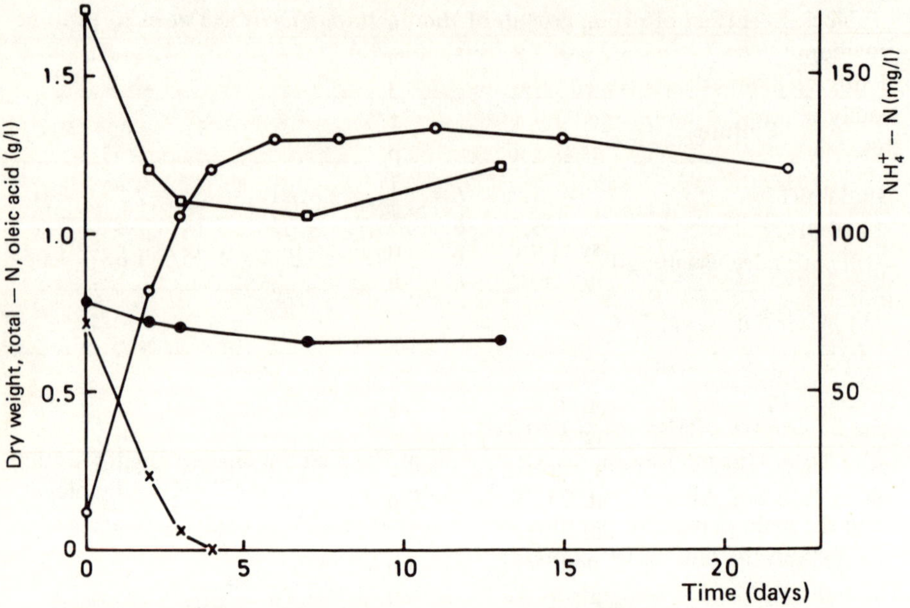

Fig. 5.2 – Growth of *Microthrix parvicella* in Tween 80/peptone medium. Oleic acid (supplied as Tween 80) x—x, total nitrogen •—•, ammonium nitrogen □—□ in medium, and dry weight o—o

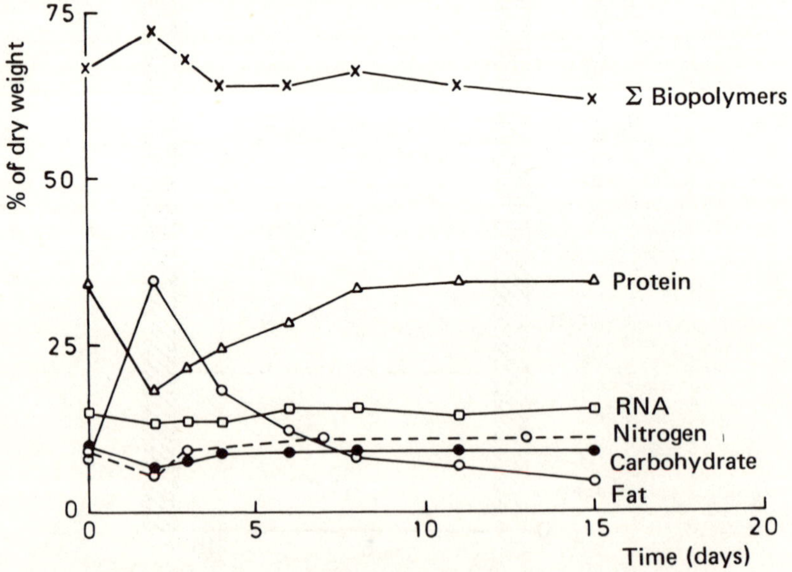

Fig. 5.3 – Composition of biomass during growth of *Microthrix parvicella* in Tween 80/peptone medium. Nitrogen o- - -o, carbohydrate •—•, RNA □—□, protein △—△, fat o—o, sum of biopolymers x—x

The changes in the composition of the medium and of the biomass during growth are given in Figs 5.2 and 5.3. It is evident that the growth of *M.parvicella* in this medium is unbalanced: the substrate (oleic acid part of Tween 80) is readily removed from the medium and partly stored as fat globules; subsequently this reserve material decreases and the protein percentage increases.

The specific growth rate of *M. parvicella* is low. No exact value can yet be given since it is unreliable to measure the specific growth rate in batch experiments, particularly in the case of unbalanced growth as it occurs in this organism. Cultivation of *M. parvicella* in continuous culture is in progress.

5.2.3 Respiration of oleic acid

To study the substrate respiration of *M. parvicella,* cells of a 12 to 15 day old batch culture were centrifuged, washed twice in water, re-suspended in water and cooled. The respiration experiments were carried out in a Yellow Springs Instruments respirometer at 25 °C in mM potassium phosphate buffer (pH $\leqslant$ 8.0) or 25 mM Tris HCl buffer (pH $>$ 8.0).

At low concentrations, neutralized oleic acid is readily oxidized at pH = 8.0. The oxidation of this substrate at the initial rate is only 10 to 20% of the theoretically calculated value (Fig. 5.4). This suggests that oleic acid is partly oxidized and then stored, followed by a slow consumption of the stored material.

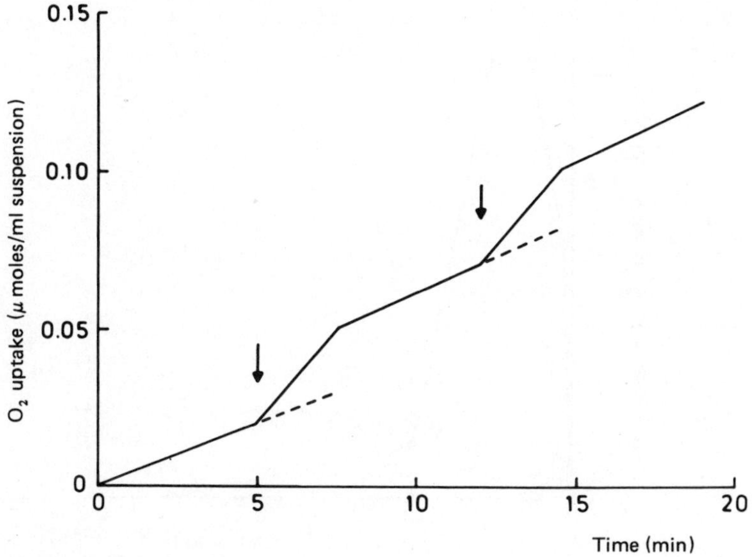

Fig. 5.4 – Schematic diagram of oleic acid respiration by washed cells of *Microthrix parvicella.* At arrow: addition of 7.5 mmoles oleic acid/ml suspension. The suspension contained 2 mg of cells, dry weight/ml

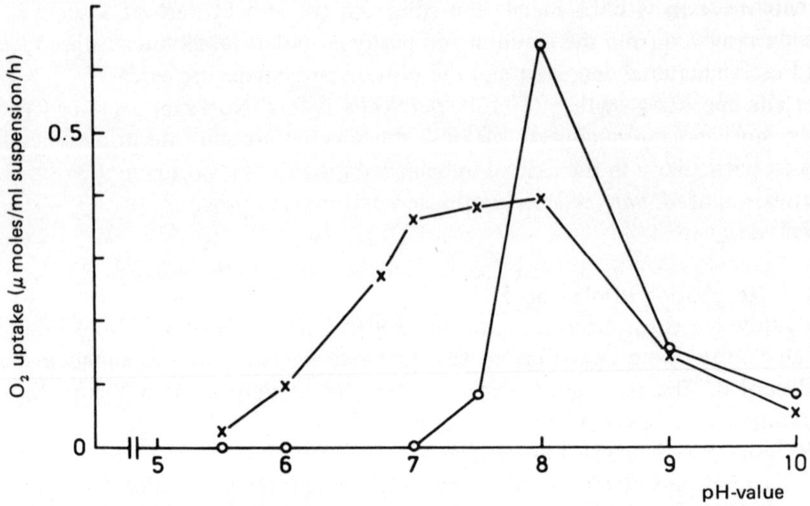

Fig. 5.5 – Effect of pH-value on oleic acid respiration by washed cells of *Microthrix parvicella.* Oleic acid concentration: 0.005 g/l x—x, 0.050 g/l o—o. The suspension contained 2 mg of cells, dry weight/ml

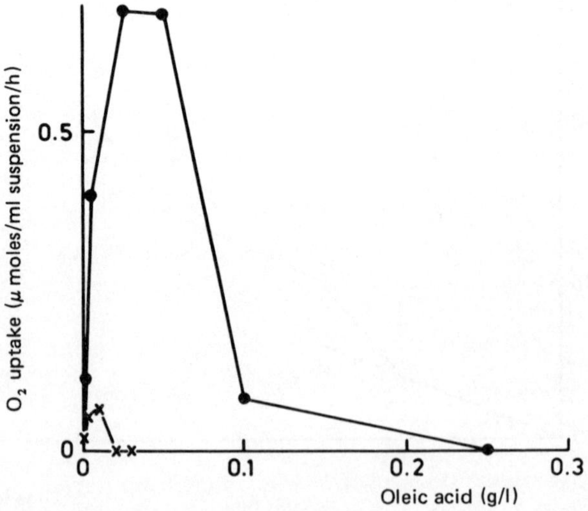

Fig. 5.6 – Effect of oleic acid concentration on the respiration by washed cells of *Microthrix parvicella.* Respiration rate: at pH 5.5 x—x, at pH = 8.0 •—•. The suspension contained 2 mg of cells, dry weight/ml

The pH dependency of respiration at two concentrations of oleic acid is shown in Fig. 5.5. At pH $\leqslant$ 7.5 the respiration at 0.05 g/l is severely inhibited whereas that at 0.005 g/l gradually decreases with pH value. The concentration of oleic acid at which respiration drops sharply depends on the pH value shown in Fig. 5.6. From the results shown in Figs 5.5 and 5.6 it may be concluded that a certain level of undissociated oleic acid causes the inhibition of the respiration. The concentration of the undissociated oleic acid is determined by the total amount of oleic acid and oleates and the pH value.

The saturation constant (equivalent to the Michaelis Menten constant) of oleic acid is within the range of 1 to 5 mg oleic acid/1 at all pH values tested.

5.3 GENERAL DISCUSSION

The high specific requirement for fatty acids is the most remarkable property of *M. parvicella.*

Eikelboom noticed that extensive growth of *M. parvicella* was not observed in low-loaded activated-sludge plants supplied with industrial waste water without a fat fraction. However, domestic sewage and many industrial waste waters contain fats and grease. The contribution of the fatty acids to the total organic matter in domestic sewage varies widely: values of 50% of the total organic carbon in suspension [10] and 22% of the suspended volatile matter [11] have been reported. Concerning the compostion of the fatty acids the reports are in better agreement. Oleic acid, stearic acid and palmitic acid account for 70% [10] to 90% [12] of the total fatty acids. The fat content of sewage increases when waste water from agro-industries (fish-processing industries, slaughter houses) and laundries is included in the sewage.

From these data it is suggested that extensive growth of *M. parvicella* can be controlled by decreasing the amount of fats in the influent. This theory has to be proved by experiments in pilot plants fed with normal and degreased waste water. Fats and grease in domestic sewage can be reduced by up to 80% using a grease trap.

The respiration and growth of *M. parvicella* are highly pH-dependent (Table 5.2 and Fig. 5.5). In a medium containing 4 g Tween 80/1 (0.8 g oleic acid/l), *M. parvicella* grows only at pH > 7 and the respiration maximum lies at pH = 8.0. The pH value of the mixed liquor in aeration tanks normally varies between 6 and 7.5 and this permits the growth of *M. parvicella* under suboptimal conditions if the concentration of free fatty acids does not exceed the inhibitory level.

High levels of free fatty acids inhibit the respiration of *M. parvicella*. At pH = 8.0 the inhibition starts at a concentration above 0.05 g oleic/l. This value is low compared to values reported by Williams and Bennet [13] for *Ps. aeruginosa.* The growth of the latter organism is not inhibited by sodium oleate at pH = 7.0 to 7.3 up to concentrations of 50 g/l.

Microthrix parvicella is very efficient in removing higher fatty acids if they are present in low concentrations. The saturation constant for oleic acid is low; however, saturation constants of other fat-consuming bacteria have not been reported.

Microthrix parvicella removes the fatty acids from the medium at a high rate. The fat is partly stored as reserve material and after exhaustion of the external substrate the reserve material is used for protein production (Fig. 5.3). This mechanism may be useful for *M. parvicella* in situations with temporarily higher fatty acid concentrations and may compensate for the low maximum growth rate.

5.4 ACKNOWLEDGEMENT

The author thanks Ir. D. H. Eikelboom of the TNO Research Institute for Environmental Hygiene, Delft, The Netherlands, for the information about the occurrence of *Microthrix parvicella* in waste water treatment plants.

5.5 REFERENCES

[1] Van Veen, W. L. Bacteriology of activated sludge, in particular the filamentous bacteria. *Antonie van Leeuwenhoek,* 1973, **39**, 189–205.

[2] Eikelboom, D. H. Filamentous organisms observed in activated sludge. *Water Research,* 1975, **9**, 365–388.

[3] Van Veen, W. L., van der Kooij, D., Geuze, E. C. W. A. and van der Vlies, A. W. Investigations on the sheathed bacterium *Haliscomenobacter hydrossis* gen.n. isolated from activated sludge. *Antonie van Leeuwenhoek,* 1973, **39**, 207–216.

[4] Van Veen, W. L., Mulder, E. G. and Deinema, M. H. The *Sphaerotilus-Leptothrix* group of bacteria. *Microbiological Reviews,* 1978, **42**, 329–356.

[5] Van Veen, W. L., Krul, J. M. and Bulder, C. J. E. A. Some growth parameters of *Haliscomenobacter hydrossis,* a bacterium occurring in bulking activated sludge. *Water Research,* 1981, in press.

[6] Mulder, E. G. and Deinema, M. H. The sheathed bacteria. In: *The Prokaryotes,* 1981, Springer Verlag, in press.

[7] Krul, J. M. Experiments with *Haliscomenobacter hydrossis* in continuous culture without and with *Zoogloea ramigera. Water Research,* 1977, **11**, 197–204.

[8] Houtmeyers, J. Relations between substrate feeding pattern and development of filamentous bacteria in activated sludge processes. *Agricultura,* 1978, **26**.

[9] Pasveer, A. A case of filamentous activated sludge. *Journal of Water Pollution Control Federation,* 1969, **41**, 1340–1352.

[10] Painter, H. A. and Viney, M. Composition of a domestic sewage. *Journal of Biochemical and Microbiological Technology and Engineering,* 1959, **1**, 143-162.

[11] Hunter, J. V. and Heukelekian, H. The composition of domestic sewage fractions. *Journal of Water Pollution Control Federation,* 1965, **37**, 1142-1163.

[12] Viswanathan, C. V., Meera Bai, B. and Pillai, S. C. Fatty matter in aerobic and anaerobic sewage sludges. *Journal of Water Pollution Control Federation,* 1962, **34**, 189-194.

[13] Williams, J. and Bennett, E. O. Biodegradation of oleates. *Journal of Water Pollution Control Federation,* 1973, **45**, 1671-1681.

Discussion

Chairman: **Dr. E. B. Pike** (Water Research Centre, Stevenage Laboratory)

Mr. I. W. Devenish (Southern Water Authority, UK) opened the discussion by commenting that it would be difficult to control the amount of long-chain fatty acids entering the sewage treatment works and he asked the author what he would recommened to be done with the fats were they to be removed from the influent sewage. In reply **Mr. Slijkhuis** suggested that if the fatty acids were removed before aeration they could perhaps be taken direct to a digestion plant or to other types of further treatment.

Referring to the graph relating uptake by *Microthrix parvicella* with pH value, **Professor P. Grau** (Institute of Chemical Technology, Prague, Czechoslovakia) said that the results suggested that if the fatty-acid concentration could be controlled to 0.05 g/l and pH value to 7 then the respiration rate, and hence growth rate of *Microthrix parvicella,* would be fairly low. He asked the author to comment on this approach as a possible control method. **Mr. Slijkhuis** did not think that this would be a satisfactory method of control. He suggested that during treatment the fatty acids would be partly removed from sewage by other bacteria so that the concentration of metabolic components for *Microthrix parvicella* would reach a sufficiently low level for growth. He referred to experiments with oxidation ditches where *Microthrix parvicella* were observed to grow well in mixed liquor with a pH value of about 6.8 and a low concentration of fatty acids. **Professor Grau** suggested that the transport mechanism for uptake of fatty acids in the bacteria was controlled. **Mr. Slijkhuis** said that another explanation for the effect could be a reversal of activation by free undissociated oleic acids. Oleic acid was not a soluble substrate and it was difficult to determine the effect of it on cell membranes, but he suggested that it might act as an inhibitor of the respiratory chain. **Professor D. Jenkins** (University of California, Berkeley, USA) asked the author what the Pk value was for oleic acid. **Mr. Slijkhuis** explained the different states of dissociation of oleic acid, the pure acid was insoluble, the salt was slightly soluble and gave a finely dispersed colloidal suspension at a pH value of 8. In the literature only the Pk values of fatty acids up to capric (C_{10}) were available. **Dr. N. Davis** (University of York, UK) commenting on Mr. Slijkhuis's work felt that the study had been very worthwhile. He was not convinced that it would be possible to get rid of *Microthrix parvicella* from activated sludge by physiological means or by the other methods Mr. Slijkhuis had suggested. He felt that it was important to study the micro-organism in mixed culture, as well as in pure culture, because the behaviour might be different in a mixed-culture system. He suggested that the results Mr. Slijkhuis had obtained were not particularly practical in terms of actual control of sludge settleability problems, but the approach of looking closely at the metabolism of a particular micro-organism was perhaps the only way in which an overall solution to the bulking problem could be achieved. **Mr. Slijkhuis** agreed with Dr. Davis's comment, and said that he hoped to perform pilot-plant studies where one plant would treat 'normal' sewage and a second plant would treat sewage containing a decreased amount of fat to see if the pure culture experiments were verified. Another experiment he hoped to carry out in the laboratory would be to feed one plant with peptone and glucose and the other with peptone and oleic acid and compare the performance of both plants under conditions of low sludge loading. He agreed that these tests were needed to confirm the results he had presented.

Professor H. Verachtert (University of Louvain, Belgium) asked about the effectiveness of other fatty acids as growth requirements for *Microthrix parvicella.* In reply **Mr. Slijkhuis** said that oleic acids could be utilized at low concentrations, but at high concentrations were inhibitory to *Microthrix parvicella.* Stearic and palmitic acids could also be used as carbon sources when fed in combination with Tween 80 but lower fatty acids were less readily taken up. Mr. Slijkhuis thought that the degradation of oleic and other long chain fatty acids was by beta oxidation and so could not explain why the bacteria could not grow on lower fatty acids. He thought this might possibly be due to the enzymatic system. **Professor P. Grau** asked how the author would account for the peptone not being taken up into the cells. **Mr. Slijkhuis** confirmed his findings that peptone was not used as a carbon source and only slightly utilized as a nitrogen source (80% of all the nitrogen appearing in the cells derived from ammoniacal nitrogen). Of the carbon only 4 to 5% of the total carbon was taken from peptone, all the rest was from the oleic acid part of the Tween 80 molecule.

Professor Jenkins asked the author whether he considered that *Microthrix parvicella* could be controlled in activated-sludge systems by using an anoxic mixing zone prior to aeration. In reply **Mr. Slijkhuis** said that the results from the laboratory trials did not suggest that this would necessarily control *Microthrix parvicella.*

General Discussion – Microbiological aspects

Chairman: **Dr. E. B. Pike** (Water Research Centre, Stevenage Laboratory)

Dr. E. B. Pike (Water Research Centre) opened the general discussion by describing two approaches to the control of sludge bulking, firstly the need to maintain bacterial diversity and not let filamentous micro-organisms predominate and secondly, the control of specific nutrients and nutrient concentrations. **Mr. D. H. Eikelboom** said that if there was a small diversity of species and one species should decline then there could be problems with the efficiency of treatment. He felt that the correct strategy was to limit the percentage of filaments in the population, for example by the control of nutrient concentrations.

Mr. A. G. Boon (Water Research Centre) referred to Dr. Green's paper (Chapter 4 of this book) where it was said that growth of *Microthrix parvicella* did not cause bulking problems at Palmersford Sewage Treatment Works (STW) and that it was other bacteria that were responsible. He referred to Mr. Eikelboom's paper (Chapter 3 of this book) where *Microthrix parvicella* had been described as one of the most important causative filaments of sludge bulking. **Mr. Eikelboom** pointed out that his work had referred to lowly loaded plants. **Mr. Boon** continued by asking why one species of bacteria should have a major effect on sludge settleability in one plant and not in another. **Mr. R. A. Varley** (Wessex Water Authority, UK) said that at Holdenhurst STW they had found that *Microthrix parvicella* could exist to a large extent in one part of the plant, and cause problems with sludge settleability, and to a lesser extent after an anoxic zone, and cause no problems with poor sludge settleability during subsequent treatment. He suggested that the effect of filamentous micro-organisms on sludge settleability was related to the actual numbers of filaments present. **Mr. Boon** continued by saying that large numbers of *Microthrix parvicella* in a low loaded plant had apparently caused problems with sludge settleability. However, a large number found in a conventionally loaded plant did not cause poor sludge settleability. He asked whether there was a link between *Microthrix parvicella* and sludge settleability.

Replying to Mr. Boon, **Mr. Eikelboom** said that he found in his original experiments that there was usually a good relationship between the extent of filamentous growth and Sludge Volume Index (SVI) although when sludge loading was increased the relationship was not always so clear. He asked Dr. Green how the identification of filaments was made. **Dr. G. P. Green** replied that the filamentous micro-organisms described in his paper (Chapter 4) were identified using the key devised by Mr. Eikelboom.

Mr. D. J. Critchard (Thames Water Authority, UK) referred to the comment of Dr. Davis during the discussion of the previous chapter when he had suggested that *Microthrix parvicella* might have different growth characteristics in a mixed culture as opposed to a pure culture. He suggested that the differences in effect of a micro-organism in an oxidation ditch compared with a conventional activated-sludge plant might be due to the different conditions leading to two different types of community in the activated sludge. In other words it might be other factors, or other micro-organisms, which affected the characteristics of *Microthrix parvicella* in the conventional plant. **Mr. H. Slijkhuis** said that the effect would not be due to a difference in density as the density of the *Microthrix parvicella* would be expected to be lower in the highly-loaded plant due to the carbon reserves of fatty acids.

Commenting on the relevance of measuring the abundance of filamentous micro-organisms, **Dr. K. C. Thompson** (Yorkshire Water Authority, UK) said that there did seem to be doubt about the correlation of abundance of filamentous micro-organisms with settleability. However, even assuming that there was a significant correlation he suggested that it would be easier to do an SDI test than to measure the Filament Index. He asked the authors to comment on the point of doing routine measurements of Filament Index and asked what the result would be used for. **Dr. G. P. Green** replied to the question in two parts. Firstly, he considered that the results he had obtained during his experiments to measure Filament Index were of interest to research workers and would lead to further understanding of the phenomenon of bulking. For example at Palmersford STW it had been found that only one species of filamentous micro-organism was related to the sludge bulking and it had also been shown that the addition of sodium hydroxide had reduced the abundance of the dominant species of micro-organism while at the same time improving the settleability. Secondly, apart from the value of the results for research purposes, he suggested that the results would give a sewage treatment work's operator a very early warning of a developing bulking problem in an activated-sludge plant.

Dr. Thompson mentioned work at Yorkshire Water Authority where, with two separate sewage treatment works operated in parallel, one plant bulked and the other did not. He commented that this kind of result made a philosophy of treatment difficult. **Mr. Eikelboom** commented that the activated-sludge system was generally very unbalanced and only a slight difference in operating conditions could radically alter the population of bacteria within the sludge. He agreed with

Dr. Green's remarks that regular microscopic examination gave an early warning of bulking sludge development.

Mr. P. H. Osmond (Southern Water Authority, UK) also had experience of two plants operating in parallel, one using diffused air and the other mechanical aeration. The mechanically aerated plant had always had a problem with bulking. He supposed this could be due to a difference in mixing characteristics. **Mr. I. D. Keegan** (British Sugar Corporation) said that his company had more success with diffused-air plants than with mechanical aeration. They had found that the use of compressed air had increased the operating temperature and better results, with a lower SSVI of activated sludge, had been achieved than with the mechanical-aeration system. **Professor D. Jenkins** (University of California, Berkeley, USA) confirmed that better performance of diffused-air plants had also been found in the USA.

Mr. G. McBrearty (Thames Water Authority, UK) asked how the filamentous bacteria developed in the activated sludge and whether their presence in the floc conferred any mutual benefit to the floc or to the filament. In reply **Mr. Eikelboom** said that the filamentous bacteria seemed to develop in the liquid phase and the floc would form around the filament. By examining the sludge microscopically unicellular bacteria could be seen attaching to single filaments and it would be around these bacteria that the new floc would be formed. In reply to Mr. McBrearty's second question, Mr. Eikelboom said that to a certain extent filamentous growth in a floc gave some protection and strength to the structure of the floc. Systems containing some filamentous bacteria usually had a less frequent occurrence of pin-point floc in the effluent.

Mr. J. F. Wright (Welsh Water Authority, UK), referring to the two plants containing *Microthrix parvicella* of which only one had bulking problems, commented that it was rather simplistic to suggest that one factor, for example the presence or absence of *Microthrix parvicella,* was responsible for the performance of any process plant. In sewage treatment there would not only be a variety of bacteria entering the plant, but also variations in temperature, mixing characteristics and composition of feed. He felt there was insufficient knowledge to look at the causes of sludge bulking in particular instances and then extrapolate generalizations which included all sewage treatment plants. Following on from Mr. Wright's comments, **Mr. A. C. Morriss** (Thames Water Authority, UK) said that each activated-sludge plant would presumably reach its own equilibrium. He asked whether any one biological equilibrium could be equated to any other. In reply **Dr. Pike** referred to studies of activated sludge receiving toxic shock. During the recovery of the sludge there would be an immediate reduction in the diversity of species present in the community, so that only the tolerant species would be left. As the sludge began to settle down the less tolerant species would become re-established, with the community perhaps returning to the original structure it had before the shock. He suggested that the microbiological structure of a sludge under steady operation was fairly random and depended on the local conditions at any particular time. There was a constant inoculation of the mixed liquor with bacteria in the sewage and most of these bacteria would be adsorbed onto the flocs and then be inactivated. He suspected that much of the particulate organic matter present in sewage that was adsorbed onto the flocs was probably bacteria.

Mr. E. I. Clark (North West Water Authority, UK) referred to Mr. Eikelboom's paper (Chapter 3) and presumed that the comments on the predominant filaments in oxidation ditches were based on results from a large number of plants. He asked for Mr. Eikelboom's comments, for the benefit of the UK, on what proportion of these oxidation ditches received unsettled sewage and whether he thought that this had relevance on the dominance of a particular filamentous micro-organism. In reply **Mr. Eikelboom** said that all the oxidation ditches studied had treated unsettled sewage. The influence of treating unsettled sewage on sludge settleability would be difficult to ascertain, a greater amount of the fat fraction would enter the aeration tank but more inorganic material would enter also. It has been said that the absence of primary sedimentation combined with a lower rate of sludge loading leads to low concentrations of filamentous micro-organisms, but in his experience 50% of all the oxidation ditches in Holland had problems with poor sludge settleability. **Mr. F. Wagner** (University of Stuttgart, West Germany) referred to experiences in Germany where it was found that, in general, works with primary sedimentation produced sludges with worse settle-ability than works without primary sedimentation.

PART 3: PROGRESS THROUGH LABORATORY AND PILOT SCALE STUDIES

Chapter 6

Biosorption and prevention of bulking sludge by means of a high floc loading

D. H. Eikelboom, TNO Research Insititute for Environmental Hygiene, The Netherlands

---

6.1 INTRODUCTION

Chudoba *et al.* [1,2,3], Heide and Pasveer [4] and Rensink [5] concluded that the excessive growth of filamentous micro-organisms is often directly related to the method of plant feeding. In 'completely-mixed' treatment plants the waste water is diluted with the total contents of the aeration basin, resulting in a low nutrient concentration for the micro-organisms. This low substrate concentration could be a decisive factor in the competition between filamentous and floc-forming bacteria in activated sludge. The excessive growth of filamentous bacteria might therefore be prevented by conversion to a process with a high loading during the mixing of sewage and sludge. To achieve this, two possibilities were mentioned:

(a) a 'plug-flow' plant configuration;

(b) a separate mixing tank for sewage and sludge.

Actually, there are no important differences between these two approaches; in both the loading is high during the initial mixing of waste water and recycled sludge. The original assumption that the substrate gradient in a 'plug-flow' plant would be decisive [1,2,5] could not be confirmed [3,4,6]. Different theoretical explanations for the phenomena observed were postulated:

- at low substrate concentrations the maximum growth rate of filamentous bacteria would exceed that of the floc forming micro-organisms. At high substrate concentrations the reverse would be the case [2];
- a separation of the exogenous and endogenous phase would favour the growth of floc-forming bacteria [6,7];
- floc-forming bacteria have a higher substrate removal capacity than the filamentous micro-organisms. However, they can only fully utilize this feature at a high loading [7].

The majority of these conclusions are based on experiments in relatively small pilot plants, usually fed with a synthetic sewage. The application of results obtained in this way to full-scale conditions is often disappointing. However, research in England [8] has shown that in 'plug-flow' plants bulking sludge occurs less often than in 'completely-mixed' plants. This supports the theory postulated. This method of bulking sludge control has also been applied with success elsewhere [9]. It should be noticed, however, that the results obtained are usually not so spectacular as those in laboratory experiments, while in several plants the excessive growth of filamentous bacteria could not be prevented in this way [10]. Obviously then, this method has limited application. It is not known whether the negative results of some experiments were caused by the mode of operation in the plants, or whether not all filamentous micro-organisms are sensitive to this method to the same extent. Therefore, in the Netherlands a research project on this subject has started. The aim of this is to investigate the effectiveness of a mixing zone/tank since conversion to 'plug-flow' conditions in an existing plant can usually only be achieved at high cost. In this chapter several aspects are discussed on the basis of preliminary results.

6.2 THE TERMS BIOSORPTION AND FLOC LOADING

The substrate removal processes occurring during the stage of mixing of recycled sludge and influent are decisive in the development of bulking sludge. Before going into this further two terms which will be used to describe this COD removal will be explained. Sewage has a very complex composition both in particle size and in range of compounds present. Practically all organic compounds supplied with the sewage are removed by the sludge flocs. Different processes are important in this removal. Firstly, there is a direct uptake by the bacterial cells, in which these compounds may be stored as reserve material. Moreover, physico-chemical processes are involved, such as adsorption by the floc and entrapment of larger particles in the open structure of the sludge floc. To describe this combination of biological and physico-chemical removal processes, which occur in activated sludge simultaneously, the term biosorption is used in this chapter. The biosorption capacity of a specific sludge can now be defined as the amount of organic material which can be taken up per unit of time. Because bacterial metabolism continues all the time, the biosorption by a certain sludge is not constant but will be dependent upon the contact time between sludge and waste water. The biosorption is calculated from the decrease of the substrate concentration (measured as BOD, COD, TOC etc.) in the liquid phase. The substrate concentration at the time $t = o$ can be calculated from the influent (I) and return sludge (RS) concentrations (COD) and flows (VOL):

$$COD_{t=o} = \frac{(COD_I \times VOL_I) + (COD_{RS} \times VOL_{RS})}{VOL_I + VOL_{RS}} \text{ (mg/l)} \quad (6.1)$$

Generally the COD_{RS} (the COD of the return sludge liquid phase) almost equals the COD of the final effluent. The biosorption at the time t = i is given by Equation 6.2.

$$\text{biosorption} = \frac{COD_{t=o} - COD_{t=i}}{MLSS} \text{ (mg COD/g MLSS)} \quad (6.2)$$

The abbreviation MLSS refers to the concentration of Mixed Liquor Suspended Solids in the aeration tank. The subscript RS is used later to indicate the concentration of Recycled Sludge.

In this chapter the term floc loading is also used. This is a measure of the level of the loading during the mixing stage. It is an instantaneous sludge loading which decreases directly after the mixing due to biosorption by the floc. The floc loading should not be mistaken for the normal expression of sludge loading. This latter variable represents the loading per day. The floc loading is calculated as follows:

$$\text{floc loading} = \frac{(COD_I \times VOL_I) + (COD_{RS} \times VOL_{RS})}{VOL_{RS} \times MLSS_{RS}} \text{ (mg COD/g MLSS)} \quad (6.3)$$

In this Equation, the COD_{RS} has also been included. Actually, this is not correct if the floc loading is used as a measure of the amount of substrate available. The final effluent always contains a certain amount of COD which is not available. If Equation 6.3 is corrected for this, it becomes:

$$\text{floc loading (II)} = \frac{(COD_I - COD_{RS})\, VOL_I}{VOL_{RS} \times MLSS_{RS}} \text{ (mg COD/g MLSS)} \quad (6.4)$$

The percentage of substrate removed can be computed in the following way:

$$\text{substrate removed} = \frac{\text{biosorption}}{\text{floc loading (II)}} \times 100 \text{ (\%)} \quad (6.5)$$

6.2.1 Experimental methods

The majority of the results were obtained from batch mixing experiments. Waste water and sludge were mixed in a cylindrical vessel (29 cm diameter, 56 cm high, liquid depth 30 cm, liquid volume 20 litres). On the inside of this vessel four baffles (each 3 cm wide) were fixed. The contents were stirred at 100 rev/min with a motor-driven paddle (diameter 9.8 cm) consisting of a circular plate with six arms (2.5 cm long by 2 cm high). Aeration was performed with air. Measurements were also carried out in existing mixing zones in treatment plants. Separation of the sludge flocs from the mixed liquor was performed by centrifugation (1500 × g for 2 min). Filtration is also possible, but this method is inferior to centrifugation (Fig. 6.1). The filtration effect not only depends on the pore size of the paper, but also on the sludge layer formed on the filter. Moreover, some sludges are difficult to filter rapidly. Of course, with low speed centrifugation 'large' particles settle also, but the results obtained approach reality much better in comparison with filtration.

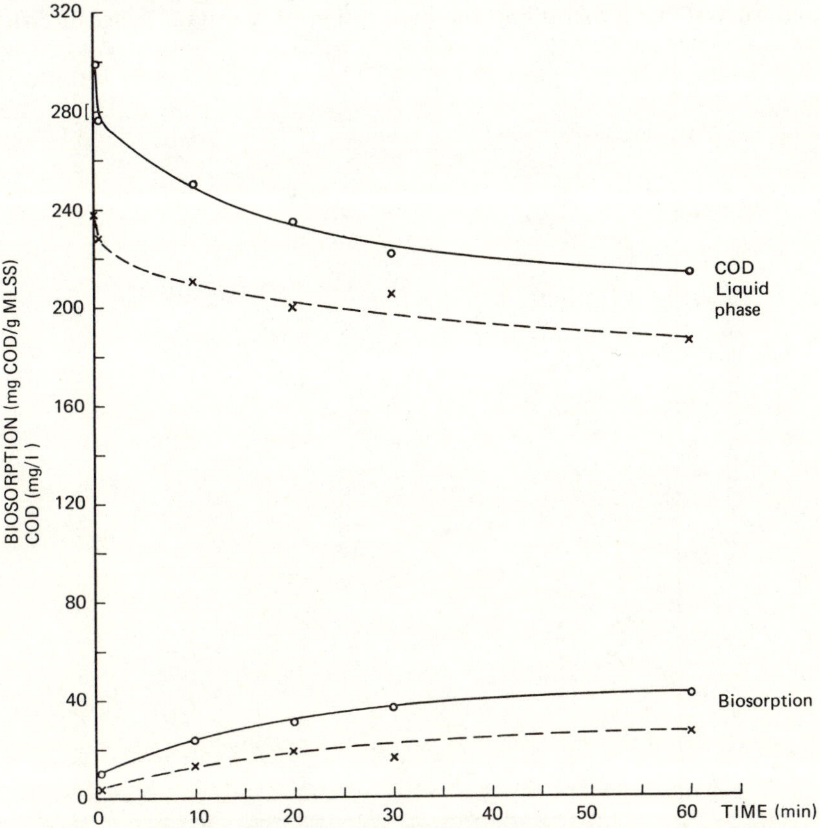

Fig. 6.1 – Comparing filtration (x——x) with centrifugation (o——o) to separate the activated-sludge flocs from the mixed liquor

6.3 BIOSORPTION BY ACTIVATED SLUDGE

It is often suggested that the organic compounds present in waste water are removed very rapidly by activated-sludge flocs. Within 5 to 10 minutes a large part of the substrate available would be fixed. Figure 6.1 shows that biosorption does not always proceed so rapidly. During this experiment, only 24 mg COD/g MLSS was taken up after 10 minutes (floc loading = 126 mg COD/g MLSS). This corresponds to 19% of the amount available. To get an impression of the substrate removal rate of sludges in full-scale plants the biosorption of 23 sludges was determined after a contact time of 10 minutes. The values measured varied from 8 to 340 mg COD/g MLSS. This corresponds to removal percentages of 13 to 80% of the substrate available. Obviously, there are large differences in biosorption capacity between different sludges. The floc loading levels during the experiments are plotted against the biosorption in Fig. 6.2. Although there is considerable scatter, probably due to the different origin of the sludges, the results seem to indicate that the loading in the mixing stage affects biosorption. To test this supposition, the biosorption of a certain sludge at different

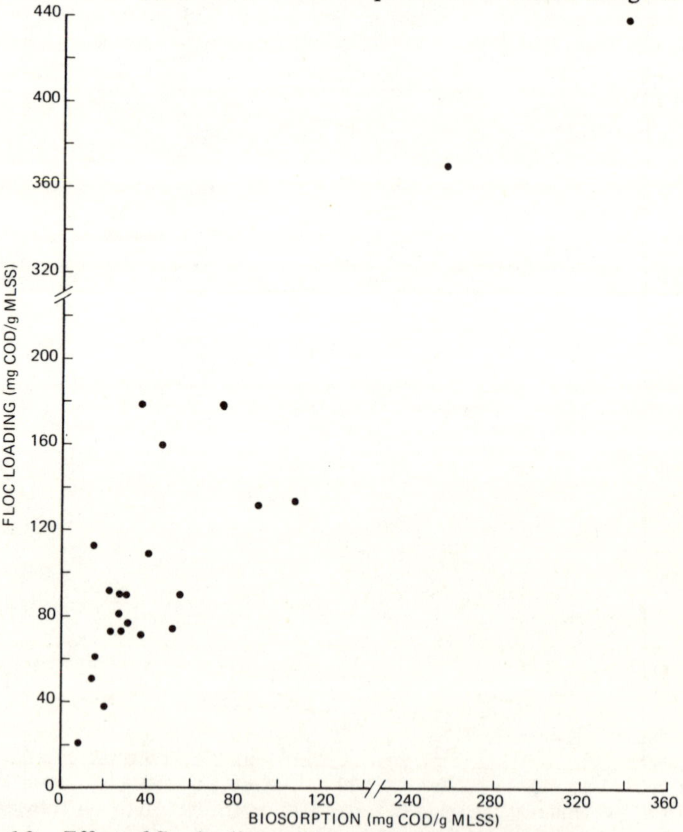

Fig. 6.2 – Effect of floc loading on biosorption by activated sludge

floc loading levels was measured. The results obtained are shown in Table 6.1 and Fig. 6.3. These data confirm the supposed relationship between floc loading and biosorption. In addition, Table 6.1 shows that the percentage of substrate removed decreases with increasing floc loading level. It is evident especially if the floc loading is not too high that biosorption is almost instantaneous, being virtually complete in one minute. Obviously, physical/chemical processes are very important [11, 12]. After this instantaneous COD removal biosorption continues considerably less rapidly because the uptake rate becomes dependent

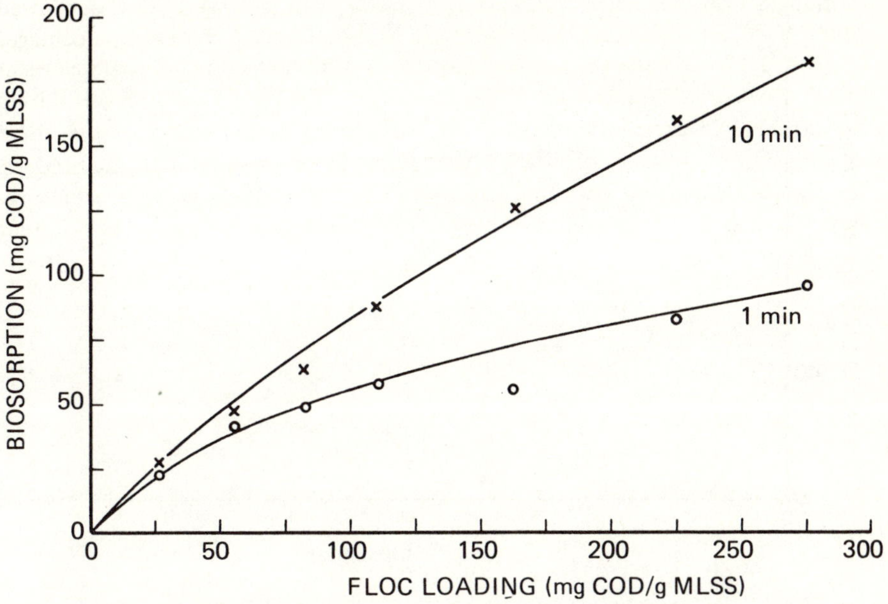

Fig. 6.3 – Effect of floc loading and contact time on biosorption by activated sludge

Table 6.1 – Effect of the floc loading on biosorption and COD removal percentages

Volume of influent (l)	Volume of sludge (l)	Floc loading (mg COD/g MLSS)	Biosorption (mg COD/g MLSS) 1 min	10 min	COD removed (%) 1 min	10 min
15.4	4.6	275	96	181	35	66
14.6	5.4	222	83	159	38	72
13.3	6.7	163	54	126	33	77
11.4	8.6	109	58	88	54	80
10.0	10.0	82	48	63	59	76
8.0	12.0	55	43	47	78	86
5.0	15.0	27	22	26	79	93

on the rate of decomposition of the compounds already fixed. Figure 6.4 shows the COD profile and the biosorption in an existing mixing zone of a treatment plant. The retention time of the sludge-sewage mixture in this zone is about 6 minutes. Within 2 minutes, about 30 mg COD/g MLSS is taken up, in the remaining 4 minutes only another 5 mg COD/g MLSS is removed. Figure 6.5 shows a graph of the biosorption values, at 1 and 10 minutes, of the 23 activated sludges mentioned previously. This figure also shows that the actual biosorption in a mixing zone consists mainly of the instantaneous COD removal.

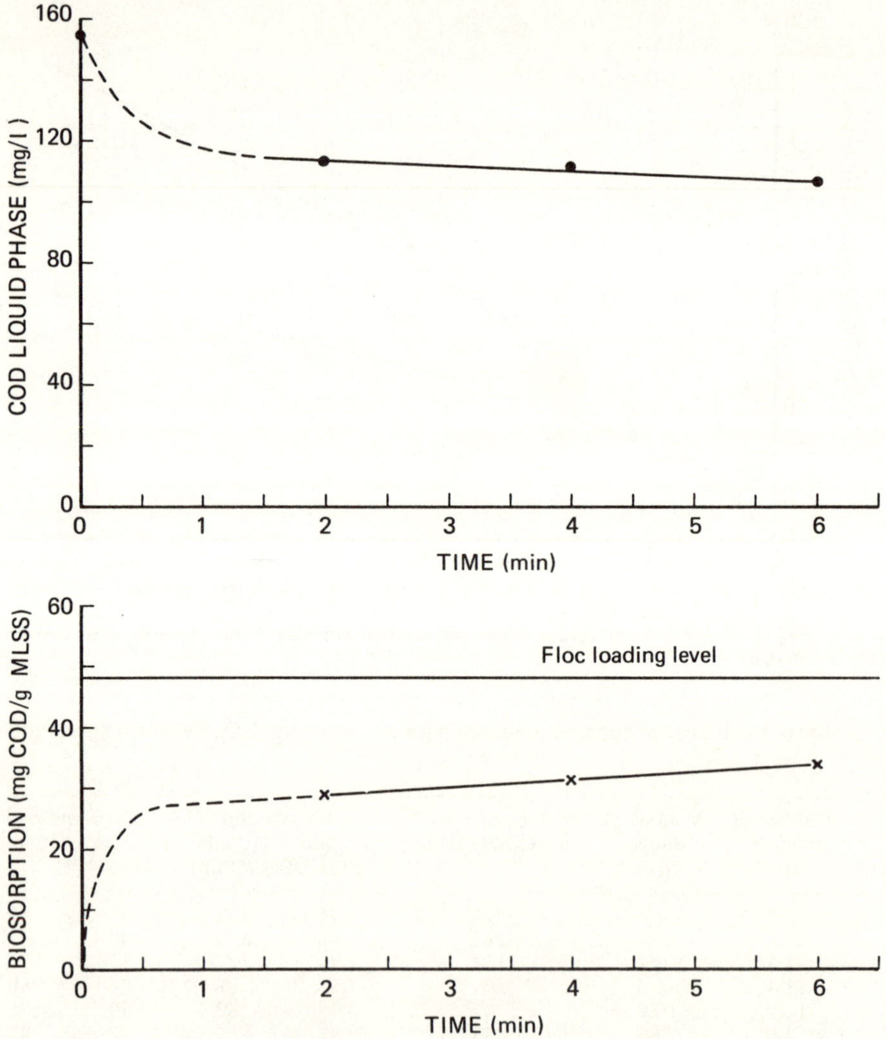

Fig. 6.4 – Course of COD reduction in liquid phase and biosorption by activated sludge in an existing mixing stage

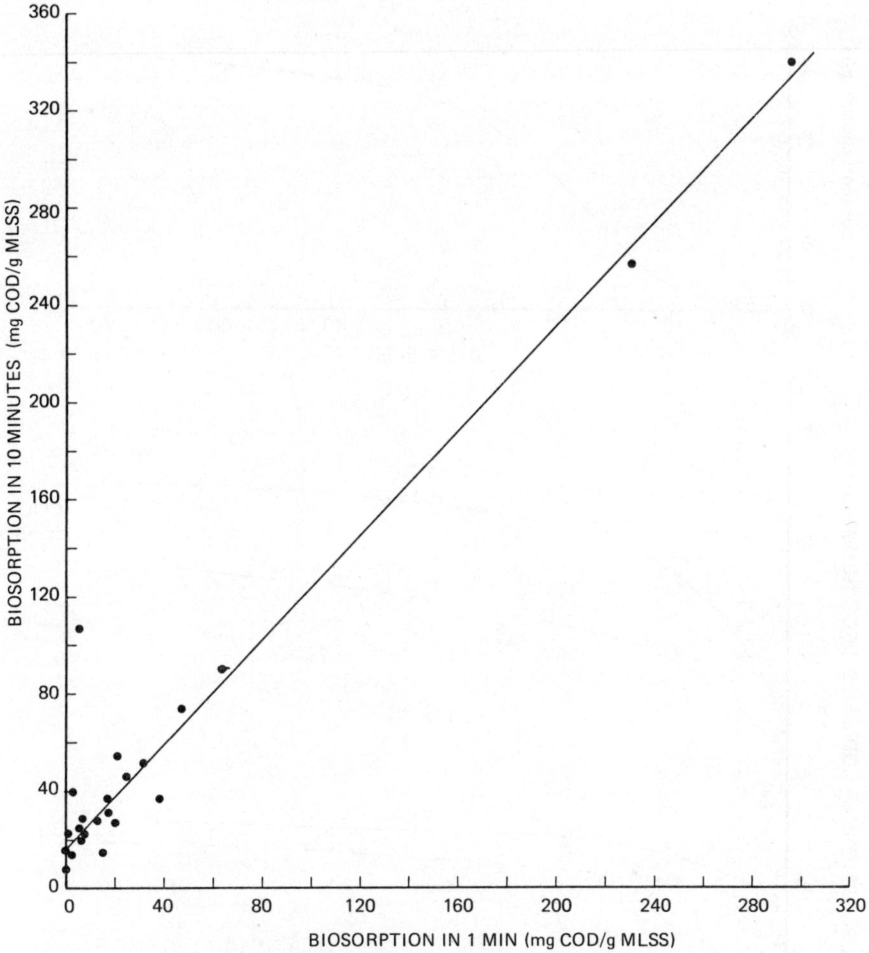

Fig. 6.5 – Relationship between the biosorption by activated sludge after a contact time of 1 and 10 minutes respectively

In the previous experiments there was sufficient oxygen available. In the next experiment nitrogen gas was blown through instead of air. In addition, a biosorption experiment was carried out with sludge and sewage of the same quality under aerobic conditions. The results are given in Fig. 6.6. The oxygen concentration increases rapidly in this experiment after a COD reduction of about 70%. In the absence of oxygen the biosorption proceeds less rapidly and the percentage removal of COD is much lower. However, the instantaneous biosorption is only slightly influenced by the absence of oxygen.

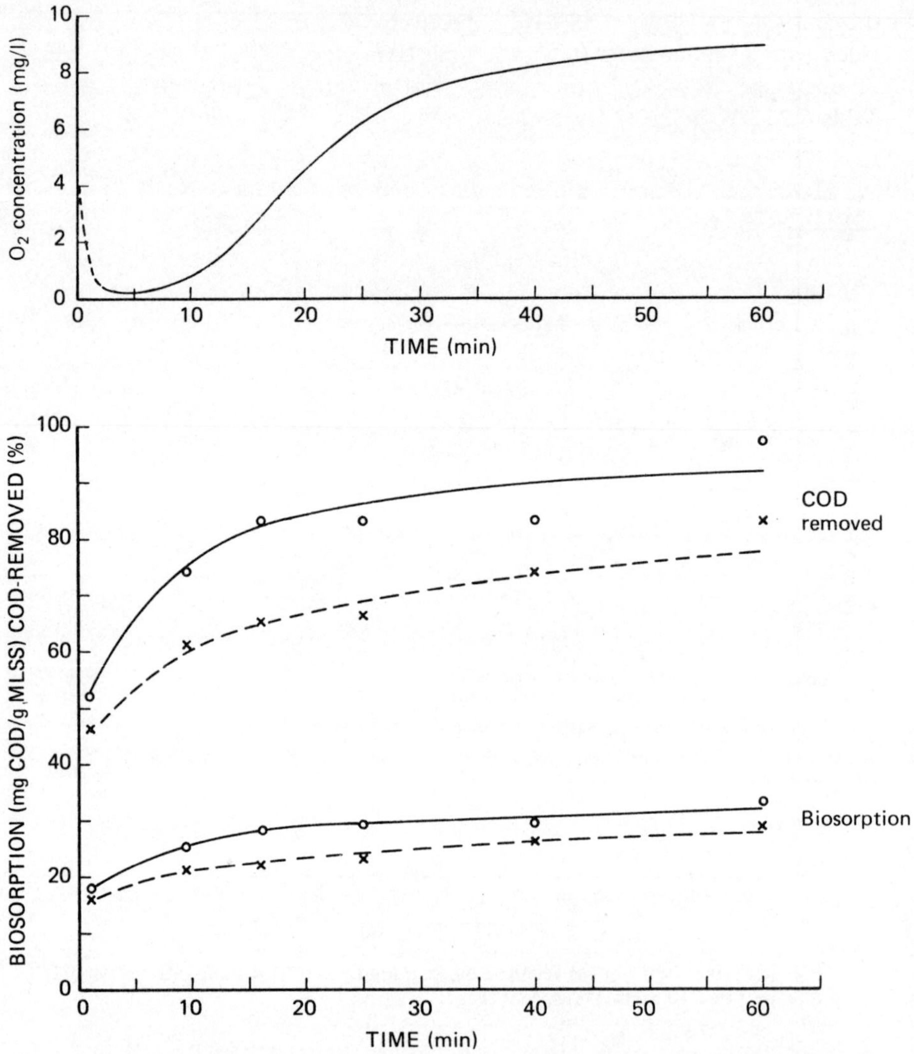

Fig. 6.6 – Effect of oxygen on biosorption by a particular sludge. o——o oxygen present (upper figure: oxygen curve) x——x oxygen absent

Organic compounds taken up by the activated-sludge floc are metabolized subsequently by bacteria. The biosorption capacity attains its original level as soon as this process is finished. The time available for this process decreases as the sludge loading of the plant increases. Thus it may be expected that the biosorption decreases as the sludge loading increases. This assumption was verified in three 'identical' pilot plants fed with domestic sewage and loaded with 0.1, 0.2 and 0.4 g COD/g MLSS day, respectively. The three pilot plants were provided with a mixing tank (retention time 10 minutes). After about four months of operation the biosorption capacity was measured. The results are given in Table 6.2 and confirm the hypothesis.

Table 6.2 – Effect of sludge loading on biosorption by activated sludge

Sludge loading (mg COD/g MLSS day)	Floc loading (mg COD/g MLSS)	Biosorption (mg COD/g MLSS)	COD removed (%)
0.1	55	30	56
0.2	52	19	36
0.4	63	15	27

It appears from other experiments that decreasing the sludge load does not always result in an increase in the biosorption. The activity of the biomass also plays a part, particularly in so-called extended aeration plants.

The time required for removing the substrate available can vary from a few minutes to many hours. However, the composition of waste water is highly heterogeneous, and a low removal rate may possibly be due to the characteristics of a particular sewage. To investigate this more closely biosorption experiments were carried out with sludge and sewage originating from two different plants. Plant A was fed with a waste water largely of non-domestic origin. This sludge had a high biosorption capacity. Sludge B originated from an oxidation ditch fed with

Table 6.3 – Biosorption experiments with sludges from two different plants

	Floc loading (mg COD/g MLSS)	Time (min)	Biosorption (mg COD/g MLSS)	Removed (%)
sludge A + sewage A	123	1	80	65
		10	104	85
		30	107	96
sludge A + sewage B	50	1	20	39
		10	46	92
		30	47	93
sludge B + sewage B	37	1	5	13
		10	13	36
		30	13	36
sludge B + sewage A	91	1	14	15
		10	22	24
		30	33	37

normal domestic sewage. Biosorption experiments were performed in which the two sludges were mixed with both sewage types. Table 6.3 shows the results. The results are influenced by the differences in floc loading, but obviously the quality of the sludge is decisive, not the type of sewage. After 10 minutes, sludge A removed about 90% of the substrate available with both kins of sewage, whereas with sludge B the percentages removed were 36% and 24% respectively.

6.4 BIOSORPTION AND CONTROL OF BULKING SLUDGE

Prevention and control of bulking sludge by means of a mixing zone can only be effective if a large part of the available substrate is actually taken up by the floc during the mixing of sewage and sludge. Only in this way can the endogenous and exogenous phases be separated clearly and hence the filamentous organisms cannot take competitive advantage of their relatively large surface area. In Table 6.4, some results of a bulking sludge control experiment in a treatment plant are shown. A mixing zone, with a retention time of approximately 6 minutes, was incorporated into this plant at the end of December 1979.

Table 6.4 – Relationship between biosorption and growth of filamentous organisms

Date	SVI (mg/l)	Category [9]	Dominating species	Biosorption (mg COD/g MLSS)
December 1979	250	4	*M. parvicella* Type 021 N *N. limicola II*	-
January 1980	175	3	*N. limicola II*	18–23
March 1980	200	2–3	*N. limicola II*	23–29
April 1980	250	4	*H. hydrossis* *M. parvicella* *N. limicola II*	2–16
June 1980	100	2–3	Type 021 N	39–42
October 1980	125	2	*M. parvicella*	50

These results demonstrate that during a period of low substrate removal by biosorption (April 1980) the settling properties of the sludge deteriorated again.

A rapid removal of organic compounds can only take place if the sludge possesses an adequate biosorption capacity. The instantaneous biosorption is particularly important because usually, from practical and financial considerations, the retention time in a mixing zone is limited. The biosorption capacity of an activated sludge appears to be variable, its extent depends on different factors, among which is the total load of the plant. The biosorption capacity decreases as the sludge loading increases. This is in agreement with the results obtained by Chudoba *et al.* [3], who observed that in their 'plug-flow' pilot plant the filamentous micro-organsims started to grow again if the sludge loading reached the value of 0.5 g BOD/g MLSS day. This decrease in biosorption is a result of the short endogenous stage. The sludge has no time actually to decompose the compounds taken up and thus restore the biosorption capacity. Temperature will probably affect the biosorption in a similar way although it has not yet been verified by experiments. It may be expected that in winter biosorption by activated sludge will be lower than in summer because a decrease in temperature will involve a less rapid decomposition of compounds taken up. Research into this temperature effect has started.

Biosorption is also greatly affected by the level of floc loading. Thus the biosorption by a certain sludge may be increased by decreasing the return sludge flow rate to the mixing zone. However, because this involves a decrease of the percentage of substrate taken up, the desired objective is not achieved (Table 6.1). For maximum removal of organic compounds, the floc loading should not be too high. A very low floc loading, however, is also unsatisfactory [13]. The filamentous organisms, which surround the floc, can take up a large percentage of the substrate supplied before this can reach the floc. We may conclude that an optimum floc loading range exists which is probably rather specific for a certain sludge quality and a certain type of sludge. In the Netherlands the best results have been obtained with a floc loading of 50 to 150 mg COD/g MLSS. However, research into the optimum floc loading level is not yet complete.

Waste water strength and flow rate usually vary over a wide range. Therefore it is almost impossible to maintain a constant floc loading. Consequently, despite the presence of a mixing stage, conditions will sometimes develop which favour the growth of filamentous bacteria. As far as we know this causes no real problems, providing that on average, the percentage removal of COD is fairly high. Some growth of filamentous micro-organisms, e.g. category 2 to 3 [9], hardly affects the Sludge Volume Index. It might be concluded from several experiments in existing plants, that about 50 to 70% of the COD supplied must be removed by the floc within 10 minutes in order to prevent an excessive growth of filamentous organisms. There are still insufficient data for a definite conclusion to be drawn.

An increase in floc loading can be achieved by conversion of the aeration basin to a 'plug-flow' pattern, or by the incorporation of a separate mixing stage. A 'plug-flow' pattern has the advantage that a sudden increase in sewage strength can be dealt with more easily (the length of the substrate gradient increases). However, separate mixing zones can be incorporated more simply into existing plants. Figure 6.7 gives some examples of this. Direct construction into the aeration basin is often preferred. It is recommended that the contents of the mixing zone are aerated. Biosorption proceeds better and more completely in the presence of sufficient oxygen and the aeration also has the side-effect of mixing the influent sewage and recycled sludge.

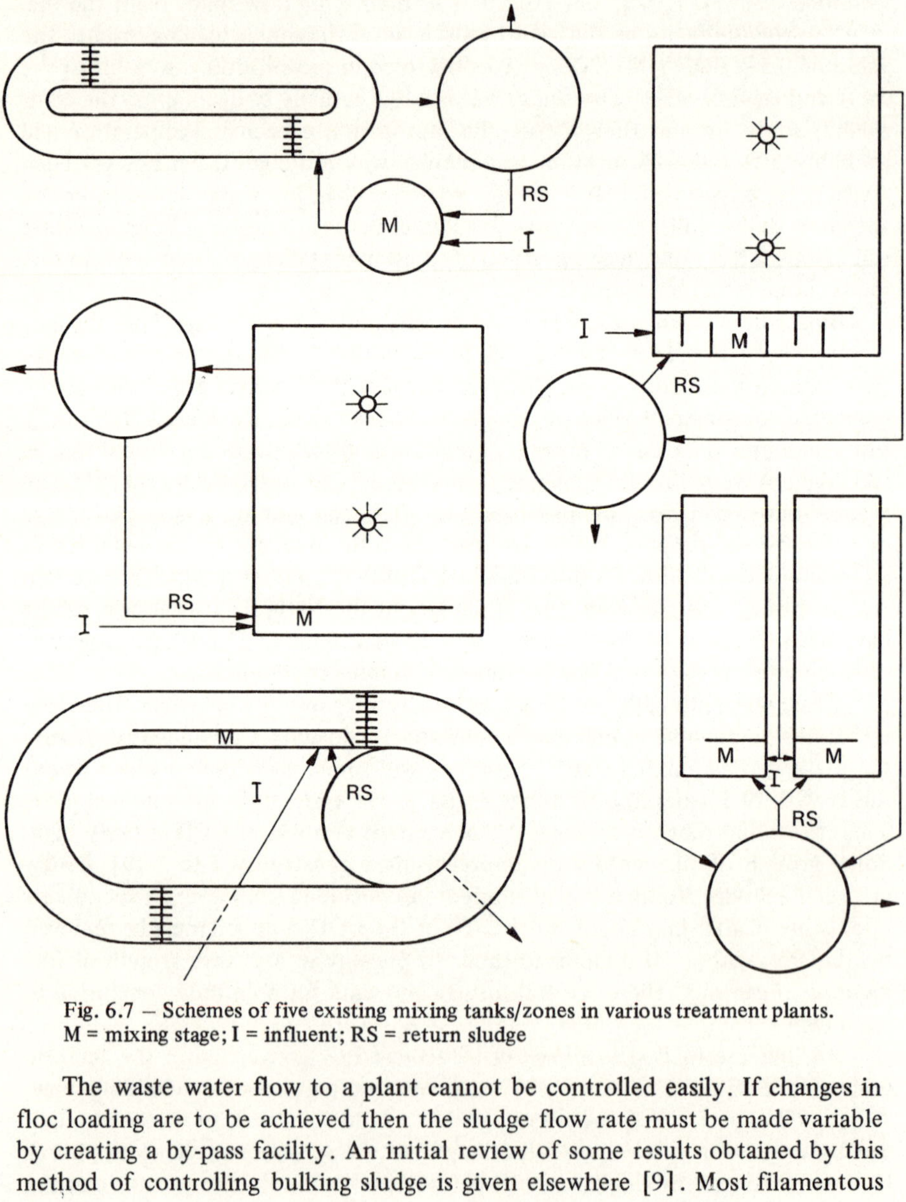

Fig. 6.7 – Schemes of five existing mixing tanks/zones in various treatment plants. M = mixing stage; I = influent; RS = return sludge

The waste water flow to a plant cannot be controlled easily. If changes in floc loading are to be achieved then the sludge flow rate must be made variable by creating a by-pass facility. An initial review of some results obtained by this method of controlling bulking sludge is given elsewhere [9]. Most filamentous micro-organisms which frequently cause bulking problems seem to be sensitive to this method. Nevertheless, this bulking sludge control method is probably not applicable in all plants. In particular, in highly-loaded plants (load > 0.4 to 0.6 g BOD/g MLSS day) hardly any improvement in the sludge settleability can be expected. It is also very doubtful whether the 0_2-concentration is decisive in this range [7, 14].

To summarize, the ultimate effect of controlling bulking sludge by a high floc loading depends mainly on the biosorption capacity of the sludge. The answer to the question of how this capacity can actually be increased has therefore become very important. An initial series of experiments investigating the effect of prolonged plant feeding by way of a contact tank has not produced the expected results. This research is being continued.

6.5 ACKNOWLEDGEMENTS

This research was partially supported by the Foundation for Applied Research for the Treatment of Wastewater (STORA).

6.6 REFERENCES

[1] Chudoba, J., Ottová, V. and Madera, V. Control of activated sludge filamentous bulking – I Effect of the hydraulic regime of degree of mixing in an aeration tank. *Water Research,* 1973, **7**, 1163-1182.

[2] Chudoba, J., Grau, P. and Ottová, V. Control of activated sludge filamentous bulking – II Selection of micro-organisms by means of a selector. *Water Research,* 1973, **7**, 1389-1406.

[3] Chudoba, J., Bláha, J. and Madera, V. Control of activated sludge filamentous bulking – III Effect of sludge loading. *Water Research,* 1974, **8**, 231-237.

[4] Heide, B. A. and Pasveer, A. Oxidation ditch: prevention and control of filamentous sludge. *H_2O*, 1974, **7**, 373-377.

[5] Rensink, J. H. New approach to preventing bulking sludge. *Journal of the Water Pollution Control Federation,* 1974, **46**, 1888-1894.

[6] Rensink, J. H., Jellama, K. and Ywema, T. The influence of the substrate gradient on the development of bulking sludge. *H_2O,* 1977, **10**, 14-16.

[7] Houtmeyers, J. Relations between substrate feeding pattern and development of filamentous bacteria in activated sludge processes. *Agricultura,* 1978, **26**, 11-35.

[8] Tomlinson, E. J. *Bulking – A Survey of Activated Sludge Plants.* Technical Report TR 35. Water Research Centre, 1976.

[9] Eikelboom, D. H. Microscopic sludge investigation and treatment plant operation. Chapter 3 of this book.

[10] Tomlinson, E. J. and Chambers, B. Methods for prevention of bulking in activated sludge. *Water Pollution Control,* 1979, **78**, 524-538.

[11] Schulze-Rettmer, R. and Yawari, T. Uber die Mechanismen der Eliminierung von Organischen Substanzen aus dem Abwasser durch Belebten Schlamm. *Z. f. Wasser und Abwasser – Farschung,* 1978, **11**, 205-209.

[12] Theophilou, J., Wolfbauer, O. and Moser, F. Die adsorptionsfahigkeit des Belebten Schlammes und Ihre Darstellung in Einem Formalkinetischen Modell des Belebtsch-ammverfahrens. *GWF (Wasser-Abwasser),* 1979, **120**, 119-125.

[13] Rensink, J. H., Voetberg, J. W. and Ywema, T. S. Y. Die Verhinderung der Entwicklung von Blähschlamm bei der Vollbiologischen Reinigung von Abwasser der Pommesfritesindustrie. *GWF (Wasser-Abwasser),* 1977, **118**, 75-79.

[14] Palm, J. C., Jenkins, D. and Parker, D. S. Relationship between organic loading, dissolved oxygen concentration and sludge settleability in the completely-mixed activated sludge process. *Journal of the Water Pollution Control Federation,* 1980, **52**, 2484-2506.

Discussion

Chairman: **G. S. Clapham**, (Wessex Water Authority, UK)

Professor D. Jenkins (University of California, Berkeley, USA) asked with reference to Fig. 6.6 if any nitrate was present in the mixing zone. He said that the figure seemed to indicate that it did not appear to make much difference to the effectiveness of the mixing zone whether there was any oxygen present or not. **Mr. Eikelboom** said that he did not think that there was any nitrate present. He said that the instantaneous biosorption was hardly influenced by the presence of oxygen. However, substrate removal due to biological activity depended on the availability of oxygen. **Professor Jenkins** said that it would be interesting to know whether any nitrate was present since it is a readily available source of oxygen. He then asked if, when calculating his floc loading, Mr. Eikelboom had used total Chemical Oxygen Demand (COD). **Mr. Eikelboom** replied that he had used only the dissolved and colloidal COD. He had separated the suspended matter from the liquid by use of a low-speed centrifuge. The numerator of the Biosorption term was the difference between the dissolved and colloidal COD in the sewage entering the zone and that in the liquor leaving the mixing zone. Mr. Eikelboom said that the COD attributed to the suspended solids was not included in the Biosorption term. **Professor Jenkins** followed up this point by saying that some of the suspended matter would be adsorbed into the activated-sludge floc in the mixing stage and might contribute significantly to the biosorption results obtained. **Mr. Eikelboom** accepted this but said that the sewage used had passed through a primary sedimentation stage and that the COD attributable to the coarse suspended solids only amounted to 10–20% of the total COD. **Professor Jenkins** then asked if Mr. Eikelboom had performed any experiments with waste waters in which all the organic matter was totally in solution. **Mr. Eikelboom** replied that he had not.

Professor P. Grau (Institute of Chemical Technology, Prague, Czechoslovakia) said that he wished to question the nomenclature used by Mr. Eikelboom, specifically the term floc loading. He said that all the other loading factors used in sewage treatment plant design included a time factor e.g. sludge loading, kg BOD/

kg MLSS day, volumetric loading kg BOD/m^3 day, surface or hydraulic loading, kg BOD/m^2 day. **Mr. Eikelboom** said that floc loading was an instantaneous form of loading. Referring to Fig. 6.5, **Professor Grau** said that the 15 mg/l of immediate removal shown on the abscissa he would call biosorption, and the later removal of COD he would call biological uptake or removal. He said that if Mr. Eikelboom were to repeat his experiment with a pure substrate, such as glucose or acetic acid, he would find a similar uptake which he did not think would be called biosorption. **Mr. Eikelboom** replied that he could not agree with Professor Grau. He had done many experiments with domestic sewage and had found that the instantaneous uptake could be very high and could be much higher than the removal due to biological activity. He said that he could probably explain Professor Grau's observations by the fact that when using a synthetic sewage an activated sludge with a much greater activity was produced than when using a domestic sewage. This would lead to totally different results from those found with domestic sewage. **Professor Grau** then showed a slide showing how the biosorption or accumulation was temperature dependent. He showed that biosorption was more temperature dependent than storage and metabolism. For all the sludges tested he had found that biosorption was highest at 20 °C.

Mr. E. I. Clark (North West Water Authority, UK) asked Mr. Eikelboom if when he was using the centrifugation technique to determine the dissolved COD in the aqueous phase he had allowed for the two minutes taken in the centrifuge in the timings recorded on the time axis of the figures. **Mr. Eikelboom** replied that the separation of sludge flocs from the liquid only took a few seconds in the centrifuge (the rest of the time being necessary for compaction of sludge pellet formed) and so the flocs were not in contact with the liquid for long during the separation and it did not need to be taken into account in the timing. **Mr. Clark** then asked Mr. Eikelboom if he could give further details about the source of the sludge used in the tests. He asked if it was derived from laboratory scale plants. He was interested to know if there were data which related the degree of biosorption with some other parameter relating to sludge quality and he asked if all the sludge was in the endogenous state and if any respiration rate data were available. **Mr. Eikelboom** replied that almost all the data were derived from full-scale activated-sludge plants treating mainly domestic sewage. He said that he did not have much data which would allow the biosorption to be related to some biological activity parameter because they did not routinely measure such a parameter. **Mr. Clark** then asked if the plants from which the data were derived were producing satisfactory effluents. **Mr. Eikelboom** replied that they were.

Mr. M. G. Booth (Wessex Water Authority, UK) followed up Mr. Clark's questions by referring to Table 6.3 where it had been shown that it was the type of sludge which was important and not the type of sewage. He asked if Mr. Eikelboom could confirm that plants A and B in Table 6.3 were full-scale plants, or were they laboratory-scale plants. **Mr. Eikelboom** said that both were full-scale plants with plant A treating a load of 30 000 population equivalents (PE) and plant B being an oxidation ditch with a load of 5000 PE. Mr Booth then said that a logical extension of Mr. Eikelboom's statement (that it was the sludge which had the major influence and not the sewage) was that it was the process configuration of the two waste water treatment plants which had the effect upon the sludge. **Mr. Eikelboom** agreed that this was so on a long term basis. He said that if the sludge of plant B which had a low biosorption capacity, was fed with the sewage of plant A, then eventually a sludge with a high biosorption capacity would be produced.

Mr. L. Rennerfelt (Ferriklor International AB, Stockholm, Sweden) asked Mr Eikelboom if there was some advantage (with respect to overcoming bulking) in adding chemical precipitants to the sewage, and if he regarded the sludge loading as being important, where would the precipitant best be added (i) into the mixing chamber, (ii) upstream of the mixing chamber, (iii) upstream return sludge, or (iv) at some position after the mixing chamber. **Mr. Eikelboom** said that he did not have a specific recommendation as to where the chemical precipitants should be added in relation to the mixing zone. If they were being added to remove phosphorus or to increase the specific gravity of the activated-sludge floc, he did not think that where they were added in relationship to the mixing zone would be important with regard to the function of the mixing zone.

Mr. J. Watkins (Yorkshire Water Authority, UK), asked Mr. Eikelboom if he thought that the growth of filamentous micro-organisms in some activated-sludge plants was a result of high interstitial COD in the mixed liquor resulting from a low biosorption capacity. **Mr. Eikelboom** replied that filamentous growth occurred when the sludge was not able to take up the sewage COD rapidly.

Professor D. Jenkins asked Mr. Eikelboom what a plant designer should do if he was in the unfortunate position of having to design a treatment plant for a new works for which no sludge existed to measure the biosorption capacity. He asked if there were any guidelines for the value of the biosorption capacity to guess at and what factors would influence the magnitude of that value. **Mr. Eikelboom** said that ideally the designer should aim at creating a floc loading in the range 50 to 150 mg COD/g MLSS. He said that the retention time required in the mixing stage could be calculated using this floc loading range. The capital cost of such a mixing stage was very low since only a low retention time was needed. **Professor Jenkins** continued by asking how the designer could guess the biosorption capacity given the characteristics of a particular waste water. **Mr. Eikelboom** replied that the only way at present was to look for a comparable existing plant and measure the biosorption capacity of the sludge in it. He said that TNO were performing research aimed at relating the biosorption capacity to the characteristics of waste waters, but he did not yet have any results he could present.

Mr. B. J. E. Hurley (Thames Water Authority, UK), said that he was very interested to see Mr. Eikelboom's definition of floc loading, but he found it to be confusing. He asked Mr. Eikelboom if the floc loading was a measure of available substrate. **Mr. Eikelboom** replied that the floc loading was a measure of the instantaneous substrate concentration and hence a measure of the nutrient concentration during the mixing stage of recycled sludge and sewage. **Mr. Hurley** continued by saying that if that was the case he found it particularly confusing since it was his understanding that COD analysis was usually carried out in the UK to determine the proportion of oxygen demand which lies between the BOD and the so-called Ultimate Oxygen Demand. He asked to what extent was Mr. Eikelboom able to distinguish between readily-degradable organic material and intractable material by COD analysis. **Mr. Eikelboom** said that he knew of only one experiment carried out by a Water Authority in Holland which had measured both COD and BOD removal and had found an almost comparable figure. He said that the results were more spectacular on a BOD basis than COD.

Mr. J. O'Neill (Yorkshire Water Authority, UK) said that a number of years ago it was general practice to have a re-aeration stage for the recycled activated-sludge but that this has now gone out of fashion. He asked Mr. Eikelboom if he thought returning to the practice of having a re-aeration stage and following this by a mixing zone prior to the normal aeration stage would result in an even greater biosorption initially. **Mr. Eikelboom** replied that he expected that it would. **Mr. O'Neill** continued by referring to the figure shown by Professor P. Grau earlier which had shown the effect of temperature upon uptake of organic material. He said that a number of years ago he had done some work on dehydrogenase activity in activated-sludge plants and there was a clear indication that there was a doubling of the activity in the summer over that in the winter.

Professor H. Verachtert (University of Louvain, Belgium) said that he thought Mr. Eikelboom should differentiate between different substrates when measuring biosorption. **Mr. Eikelboom** replied that they had not determined concentrations of specific substrates but only COD. **Professor Verachtert** continued by saying that it was important to stress the importance of the type of substrate. He said that it was important to differentiate between the soluble substrates which can be assimilated immediately, and insoluble substrates which have to be hydrolysed enzymatically outside the cell. He believed that hydrolysis of the insoluble substrates was the rate-limiting step whether or not there was biosorption. **Mr. Eikelboom** said that he believed that the insoluble substrate was bound to the floc by a physico-chemical process. **Professor Verachtert** said that although this might be so, such substrates would not necessarily be metabolized by the bacteria. **Mr. Eikelboom** agreed that perhaps the insoluble substrate took a long time to be used by the floc bacteria, but the most important thing was that the adsorbed insoluble substrates had been removed from the mixed liquor and hence were not available for the filamentous bacteria, the greatest part of which develop outside the floc in the mixed liquor.

Mr. A. J. Rachwal (Thames Water Authority, UK) said that in a typical activated-sludge plant the concentrations and amount of a substrate could vary by a factor of three over a day, but the rate of sludge recycle was usually kept constant (at least in British practice). He said that if a mixing zone was used as recommended by Mr. Eikelboom, the floc loading in it would also vary by a factor of three. He asked whether Mr. Eikelboom would recommend linking the rate of sludge recycle to the incoming flow rate, and using a factor determined by floc loading, to set the exact ratio between recycled sludge rate and incoming flow. **Mr. Eikelboom** said that it was better to work with a constant floc loading but it is not always possible to achieve this on full-scale plants. He said that it was his experience that the normal fluctuations in sewage substrate concentrations did not have much effect upon the overall plant performance. He said that plants operated in Holland with a mixing stage (with the normal variation of sewage composition and a constant sludge recycle rate) still had some filamentous bacteria present but to a much lower extent than those without the mixing stage.

Mr. L. J. Webb (PIRA, Leatherhead, Surrey, UK) followed earlier questions relating to the effect of soluble and insoluble COD removal by asking if Mr. Eikelboom had done any work to examine the effect of the molecular weight of the soluble material upon biosorption. He was particularly interested in the pure adsorption capacity of the biological solids since he thought that there would be a great deal of difference between the adsorption of low molecular weight sugars and high molecular weight compounds such as starch. **Mr. Eikelboom** agreed that this might be true but he had not done any work to examine this effect. **Mr. Webb** continued by saying that he felt that the success or otherwise of the process modification in controlling bulking would be very much dependent upon the character of the waste, particularly with respect to the soluble BOD. **Mr. Eikelboom** said in reply that all his work had been with domestic sewage and it might be that his method was not suitable for some types of industrial waste waters. **Mr. Webb** then asked Mr. Eikelboom what he meant by the terms exogenous and endogenous and did he include in the term endogenous the respiration of adsorbed materials **Mr. Eikelboom** said that by endogenous he meant the period when substrate taken up by the cell had been more or less broken down. For activated sludge it was very difficult to separate the two stages because one did not know how far substrates which are broken down are already in the cell or initially attached to the floc and then subsequently broken down.

Mr. C. B. Waller (Thames Water Authority, UK) referring to Fig. 6.3 where the instantaneous biosorption was related to the floc loading, said that presumably Mr. Eikelboom, by using the term instantaneous biosorption, was trying to differentiate between the substrate physico-chemical adsorption and the actual substrate which was metabolized by the sludge. He said that for a given sludge one might expect that the biosorption might be related to the specific surface area and so one might expect, in a graph similar to Fig. 6.3, a fast uptake followed by a plateau where the maximum physico-chemical adsorption had been reached and there was no further adsorption. He wondered whether the fact that both the one minute and ten minute biosorption curves continued increasing with floc loading mean that that maximum physico-chemical adsorption had not been reached. **Mr. Eikelboom** agreed with this analysis and said that he expected that when the floc loading was increased to a very high level, the floc would become saturated with organic compounds.

Chapter 7

Theory and practice of accumulation-regeneration approach to the control of activated sludge filamentous bulking

P. Grau, J. Chudoba, M. Dohányos, Prague Institute of Chemical Technology, Czechoslovakia

7.1 INTRODUCTION

According to the growth theory of micro-organism selection based on the relation of μ, the specific growth rate, to K_S, the saturation constant [1], the suppression of filamentous micro-organisms in mixed cultures can be achieved in a reactor with a sufficient concentration gradient of the substrate occurring along its length. This has been shown experimentally [1, 2], and it was found that a relation between the Sludge Volume Index (SVI) and the Dispersion Number of an aeration tank did exist. However, the results of further experiments [3] as well as the results published by other investigators [4-12] have indicated that the concentration gradient of the substrate is not the only factor controlling the selection of non-filamentous micro-organisms in activated sludge. It was found at higher sludge loadings [3, 4, 5, 6], that activated sludge possessed high values of the SVI even in systems with sufficient 'plug-flow' characteristics to provide high substrate concentration gradients.

It has often been stated [13, 14, 15] that systems with higher DO concentrations, (pure oxygen), promoted lower SVIs. On the other hand, no significant differences in SVIs and excess sludge production in air and pure oxygen systems have also been reported [16, 17, 18]. Our laboratory experiments support the latter findings. Four completely mixed units (see Table 7.1 units Nos 8–11) were operated under identical conditions but at different steady-state DO concentrations, namely 1 mg/l (air), 7.5 mg/l (air), 7.5 mg/l (pure oxygen), and 16.0 mg/l (pure oxygen). The activated sludges developed in all the units were filamentous and possessed SVIs of 393, 287, 279, and 270 ml/g, respectively. The most probable cause of both the lower SVI and the excess sludge production observed in pure oxygen systems, lies in a compartmentalization of the reactor rather than in a higher DO concentration. The compartmentalization leads to a lower Dispersion Number and in turn to a concentration gradient of the substrate and a selection of non-filamentous micro-organisms. Simultaneously, in the systems with prevailing non-filamentous sludge a higher proportion of the substrate removed is oxidized than in the systems with filamentous sludge. This fact follows from the findings presented in this chapter and also from those obtained by Kroiss and Ruider on Vienna sewage [19].

Another factor that might have influence on the selection of non-filamentous micro-organisms is the regeneration time of activated sludge. The term regeneration time has been used to designate a period of sludge aeration occuring after the removal of the exogenous removable substrate. The effect of regeneration time can be demonstrated on the mixed cultures developed in aeration units Nos 21, 15, and 12 (Table 7.1). They possessed values of SVI of 51, 212, and 464 ml/g, respectively. Corresponding regeneration periods were approximately 3.0, 1.0 and 0.0 hours. Sludge regeneration in these units took place in the reactors. In practice, it is usually achieved in a separate tank called the regeneration, reaeration or stabilization tank. Many investigators [8, 9, 20, 21, 22] have observed that separate sludge regeneration improved the SVI.

From this it follows that the regeneration period plays an important role in the selection of non-filamentous micro-organisms in activated sludge. The regeneration period, however, requires a modified definition as will be shown later in the text.

7.2 ACCUMULATION–REGENERATION THEORY OF MICRO-ORGANISM SELECTION IN MIXED CULTURES

The applied theory is based on the accumulation, storage and metabolism concept [20, 23, 24] rather than on the growth–death concept [22]. It is well known that some bacterial species prior to the numerical division of cells, accumulate the exogenous substrate and convert it into various kinds of endogenous accumulation and storage compounds and other cell constituents. During this period the mass of cells increases without any substantial increase in the number of cells. Proper division, associated usually with an increase in the DNA content of the cells, follows after the accumulation period. With activated sludge and glucose as a substrate, Speece *et al.* have shown that sludge biomass can increase 200 to 270 per cent prior to cell replication [24]. Generally, storage was found to be a function of the loading rate. At lower loading rates, activated sludges showed an extended time lag (5 hours) before cell replication, while at high loading rates shorter time lags occurred before the replication. It was concluded that 'in the contact stabilization process, cell replication would be expected in the stabilization tank with only substrate storage taking place in the contact tank, due to the low loading rates used'. This statement is in good agreement with that made by Gujer and Jenkins [25]: 'COD removed in the contact basin is not entirely associated with cell growth. The COD removed is associated with the activated sludge in a partially unassimilated form that has a low COD equivalent and only becomes fully assimilated in the stabilization basin'. Thus it is evident that accumulation and storage are common phenomena in activated sludge. A schematic model of the substrate transport into the cell, its accumulation inside and its transformation to storage products is shown in Fig. 7.1 [26]. The quantity of substrate transformed to storage material in a unit weight of cells prior to their replication can be designated as storage capacity (SC). From the experiments mentioned above it follows that with activated sludge micro-organisms and glucose as a substrate, the maximum value of the SC is about 2 g/g. The quantity of substrate that can be accumulated in a unit weight of cells can be designated as accumulation capacity (AC). By means of kinetic studies [26] it was shown that with activated sludge and glucose the maximum values of the AC were 0.3 to 0.4 g/g, i.e. about 20 per cent of the SC [26]. The AC is also strongly dependent on the sludge age and on the proportion of glucose in the substrate [26]. Since in waste water treatment practice the initial loadings are low (0.1 to 0.5 g/g), it seems probable that the AC rather than the total SC will play a significant role in micro-organism selection.

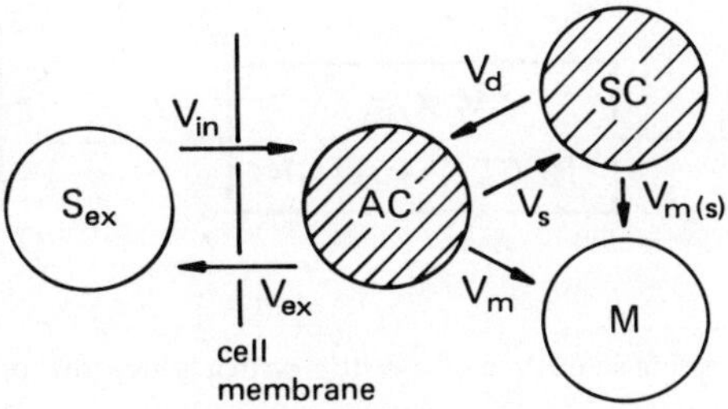

S_{ex} = extracellular substrate concentration
V_{in}, V_{ex} = rates of transport into and out of cell, respectively
V_d = rate of depolymerization of storage material
V_s = rate of substrate modification and storage
V_m = rate of substrate metabolism
$V_{m(s)}$ = rate of storage material metabolism
M = metabolic sequence.

Fig. 7.1 – Schematic model of saccharidic substrate transport into the cell, its accummulation and transformation to storage products and metabolism.

It is logical to assume that, firstly, the maximum AC and rates of accumulation are different for various species, and secondly, the available AC will change according to the regeneration period of cells. If the regeneration period is sufficient to restore fully the ACs of all species in a mixed culture, then the predominant species in the culture will be that with the highest AC and the highest rate of accumulation. Through this mechanism, a natural selection of species in activated sludge can be controlled. Examination of the literature has revealed that there is no satisfactory method of calculating the necessary regeneration period. An attempt to develop a new accumulation–regeneration model is presented in this study. The model suggested by Yakovlev and Karyukhina [27] served as a starting point.

7.3 ACCUMULATION–REGENERATION MODEL

The model will be derived and explained for a system with a 'plug-flow' pattern and separate sludge regeneration, the schematic outline of which is shown in Fig. 7.2. The model originates from the fact that activated sludge micro-organisms undergo an accumulation–regeneration cycle (feeding cycle) several times per day. The duration of one cycle is given by the total detention period of micro-organisms in the contactor and regenerator.

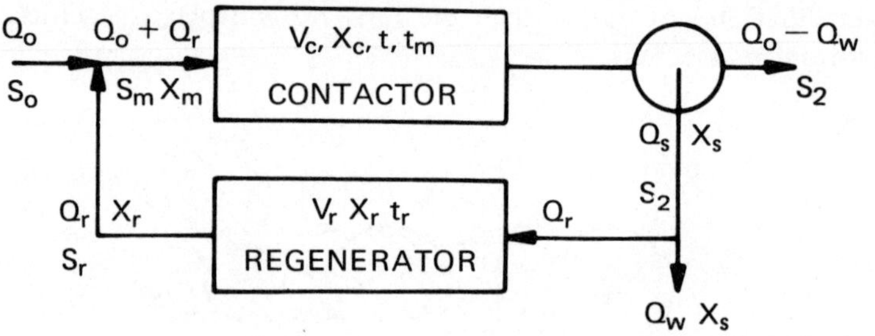

Fig. 7.2 – The schematic outline of a system with a 'plug-flow' pattern and separate regeneration.

The first key parameter in the accumulation–regeneration model (AR model) is the initial loading, L_i (see Fig. 7.2):

$$L_i = \frac{S_m}{X_m} \tag{7.1}$$

Both variables can be calculated from a material balance:

$$S_m = \frac{S_o + RS_r}{1 + R} \tag{7.2}$$

$$X_m = \frac{R}{1 + R} X_r, \tag{7.3}$$

where $R = Q_r/Q_o$.

Equation 7.1 defines the initial loading in a continuous system with the 'plug-flow' pattern. However, in order to calculate the required regeneration period, the quantity of substrate removed per unit weight of the initial Mixed Liquor Suspended Solids (MLSS) in one AR cycle must be known. It can be expressed as follows:

$$\Delta L_i = \frac{S_m - S_2}{X_m} \tag{7.4}$$

Substituting S_m from Equation 7.2 for S_m in Equation 7.4 yields:

$$\Delta L_i = \frac{S_o - S_2 + RS_r - RS_2}{(1 + R) X_m} \tag{7.5}$$

Since in most activated-sludge systems with high purification efficiencies $S_r \doteq S_2$ and $X_m \doteq X_c$, Equation 7.5 reduces to:

$$\Delta L_i = \frac{S_o - S_2}{(1 + R) X_c} \tag{7.6}$$

Equation 7.6 determines the milligrams of substrate removed by one gram of the MLSS in one AR cycle. Of the total amount of substrate removed, a part is oxidized and a part accumulated and stored. As the regeneration continues the accumulated and stored material is further oxidized in order to supply the cells with energy and is partly converted to new cells.

If p is a proportion of the substrate that should be oxidized in order to restore fully the accumulation capacity, then the specific quantity of the substrate oxidized during the regeneration period is:

$$(\Delta L_i)_{ox} = p\Delta L_i = \frac{S_o - S_2}{(1 \div R) X_c} \quad p \tag{7.7}$$

The regeneration period necessary for this oxidation will evidently depend on an average specific respiration rate, $\bar{r}$, in the whole AR system:

$$t_n = \frac{(\Delta L_i)_{ox}}{\bar{r}} \tag{7.8}$$

After being brought into contact with the substrate at the inlet end of a contactor, micro-organisms start to oxidize the substrate and so the sludge regeneration immediately proceeds in the contactor. Consequently, in the AR model the regeneration time is defined as a total period of activated-sludge aeration in one AR cycle:

$$t_s = t_m + t_r \tag{7.9}$$

The second key parameter in the AR model is the average specific heterotrophic respiration rate, $\bar{r}$. Its correct calculation is very important. Values of $\bar{r}$ for different combinations of the AR system can be calculated as follows:

$$\bar{r} = \frac{r_r t_r + \int_0^{t_m} r(t)dt}{t_s} \tag{7.10}$$

If $t_r = 0$, then $t_s = t_m$ and $\bar{r}$ represents an average respiration rate in a 'completely-mixed' reactor or in a reactor with 'plug-flow' pattern. Thus the AR model is also applicable to any activated-sludge system without separate sludge regeneration.

It is hypothesized that activated sludge with a minimum content of filamentous micro-organisms and consequently with low values of the SVI, can be obtained under the condition:

$$t_s \geqslant t_n \tag{7.11}$$

However, in order to calculate the necessary regeneration period, t_n, it is important to know the proportion, p, of the substrate that should be oxidized during this period. Combining Equations 7.7 and 7.8 leads to:

$$p = \frac{\bar{r}\, t_n X_c (1 + R)}{S_o - S_2} \tag{7.12}$$

It can be demonstrated easily that the right-hand side of Equation 7.12 is equal to the ratio of average respiration rate to sludge loading rate L:

$$p = \frac{\bar{r}\, \bar{X}\, \bar{t}}{S_o - S_2} = \frac{\bar{r}}{L}, \tag{7.13}$$

where $\bar{X}$ is the average concentration of activated sludge in the AR system and $\bar{t}$ is the average detention time of waste water in the system. For AR systems with separate sludge regeneration both variables are defined as follows:

$$\bar{X} = \alpha' X_c + \beta' X_r, \tag{7.14}$$

where $\alpha' = \dfrac{V_c}{V_c + V_r}$, $\beta' = \dfrac{V_r}{V_c + V_r}$

$$\bar{t} = \frac{V_c + V_r}{Q_o} \tag{7.15}$$

The following relation holds between t_s and $\bar{t}$:

$$t_s = \bar{t}\left[\frac{\alpha'}{1+R} + \frac{\beta'}{R}\right] \tag{7.16}$$

For AR systems without separate sludge regeneration, Equation 7.12 reduces to:

$$p = \frac{\bar{r}\,t\,X}{S_o - S_2} = \frac{\bar{r}}{L}, \tag{7.17}$$

where $t = t_m\,(1 + R)$, $t_n = t_m$, $X = X_c$

As the higher regeneration periods result in a higher proportion of the substrate oxidized, it is assumed that a relation between the SVI and p exists.

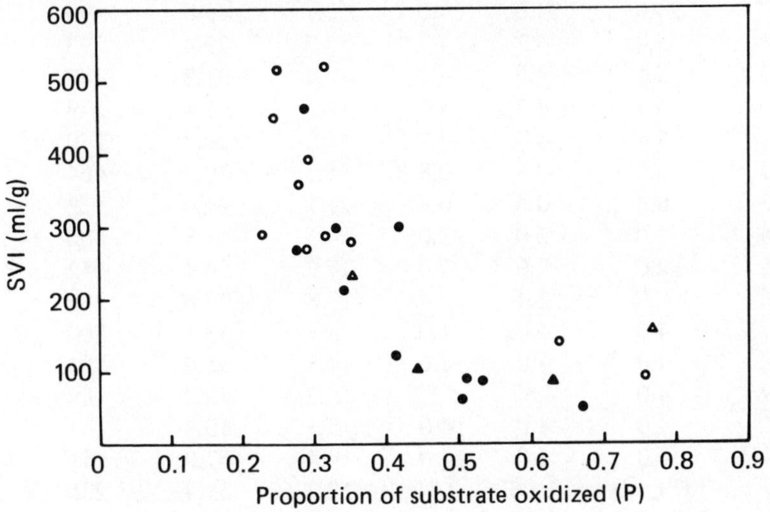

Fig. 7.3 – Relationship between SVI and proportion of substrate oxidized, o – 'completely-mixed' systems (CMS); • – system with 'plug-flow' pattern (PFPS); Δ – CMS with separate sludge regeneration; ▲ – PFPS with separate sludge regeneration.

7.4 EXPERIMENTS

All experiments were carried out in laboratory activated-sludge units having total volumes from 0.8 to 24 litres. A summary of the operational parameters of individual units is given in Table 7.1. Aeration units Nos 1 to 11 were 'completely-mixed' reactors, units Nos 12 to 21 were reactors with the 'plug-flow' patterns of different Dispersion Numbers (0.033 to 1.06), and units Nos 22 to 25 were 'contact-stabilization' systems with 'completely-mixed' regenerators. The contactors were reactors with 'complete-mixing' (Nos 22 and 23), or 'plug-flow' pattern (Nos 24 and 25). The substrate used in all experiments was a mixture of starch (300 mg/l) and peptone (300 mg/l) dissolved in tap water and balanced

Table 7.1 – Summary of the operational parameters and results obtained

Unit No	Total volume (l)	t_m (h)	t_s (h)	$\bar{r}$ (mg/gh)	p (%)	SVI (ml/g)	A (d)
1	0.8	0.8	0.8	29.0	22.9	293	2.0
2	0.8	0.8	0.8	36.9	28.1	358	3.0
3	2.0	2.0	2.0	23.0	24.6	450	2.0
4	2.0	2.0	2.0	27.4	31.8	523	3.0
5	4.0	4.0	4.0	11.0	25.1	517	3.0
6	6.0	7.1	7.1	24.7	63.7	141	3.6
7	24.0	24.0	24.0	2.7	75.8	95	125.0
8	5.4	4.7	4.7	13.7	29.4	393	3.2
9	5.4	4.7	4.7	12.4	31.8	287	4.6
10	5.4	4.7	4.7	12.1	35.4	279	4.7
11	5.4	4.7	4.7	11.2	29.3	270	4.7
12	0.8	0.8	0.8	55.3	29.0	464	2.0
13	0.8	0.8	0.8	21.1	41.4	122	3.0
14	2.0	2.0	2.0	44.5	41.8	300	2.0
15	2.0	2.0	2.0	17.2	34.4	212	3.0
16	2.0	2.4	2.4	13.8	27.6	270	3.7
17	4.0	4.0	4.0	8.7	33.1	300	4.8
18	4.0	4.0	4.0	14.8	51.0	91	4.7
19	4.0	4.57	4.57	16.2	53.2	89	4.5
20	4.0	4.0	4.0	15.3	50.4	61	4.5
21	4.0	4.0	4.0	17.4	67.0	51	4.5
22	1.0 + 1.0	1.16	3.96	12.2	35.4	235	3.7
23	3.0 + 3.0	3.57	12.32	18.5	76.8	157	6.3
24	1.0 + 1.0	1.25	4.55	14.6	44.3	106	4.4
25	0.8 + 1.2	1.05	5.67	17.3	62.9	86	9.0

with inorganic nutrients. Influent values of COD and BOD_5 were 640 and 300 mg/l, respectively. Values of the SVI were determined in 100 ml cylinders without stirring. All respiration rates were corrected for nitrification. Values of p were calculated according to Equations 7.13 and 7.17.

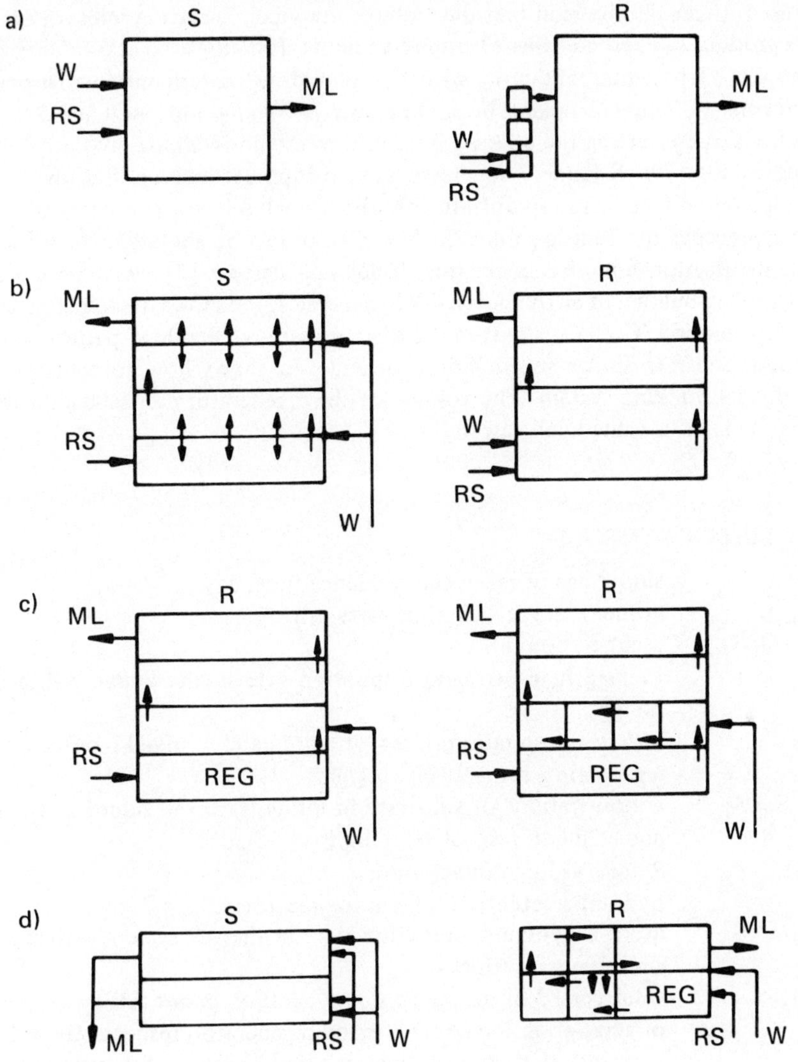

Fig. 7.4 – Schematic outline of aeration systems supporting (S) and repressing (R) filamentous bulking. W – waste water, RS – return sludge, ML – mixed liquor, REG – regenerator.

7.5 RESULTS AND DISCUSSION

The results are summarized in Table 7.1 and in Fig. 7.3. This figure demonstrates clearly a relation between the SVI and the proportion of the substrate oxidized. It can be seen that the mixed cultures with low values of the SVI can be obtained only in such systems in which at least 50 per cent of the substrate removed was oxidized. For safe design, however, the p-value of 0.65 rather than 0.5 should be used. It can also be seen that the 'completely-mixed' aeration tanks or contactors produced mixed cultures with the SVI above 100 ml/g even at p-values higher than 0.6. Hence, the contactors with the 'plug-flow' pattern and low Dispersion Numbers are more efficient in suppression of filamentous bulking. Practical methods of converting the bulking systems into the non-bulking ones are summarized in Fig. 7.4. Scheme 7.4a shows how to suppress bulking in a low loaded 'completely-mixed' aeration unit. In this case a small selector consisting of 3 to 5 units precedes the aeration unit. Such a selector can be simultaneously used as a denitrification unit. Realizing this, Heide and Pasveer [7] were successful in suppressing bulking in an oxidation ditch. Scheme 7.4bc shows the transformation of step aeration (Gould system) to the systems with a 'plug-flow' pattern with or without separate sludge regeneration. Scheme 4d shows how to transform an overloaded bulking system. The volume of the regenerator has to be adjusted to the calculated t_s value in all cases.

7.6 LIST OF SYMBOLS

A	sludge age or mean cell residence time, day
Q_o,Q_r	influent and recycle flow rates, l/h
$R = Q_r/Q_o$	recirculation ratio
$\bar{r}$	average heterotrophic respiration rate in the whole AR system, mg/gh
r_c,r_r	Specific respiration rates in completely mixed contactor and regenerator, respectively, mg/gh
S_o,S_m,S_2	concentrations of substrate in influent, mixed liquor at the inlet and effluent, respectively, mg/l
SVI	Sludge Volume Index, ml/g
t	hydraulic detention time in the reactor, h
$\bar{t}$	average hydraulic detention time in the AR system with separate sludge regeneration, h
t_s,t_m,t_r	total period of activated sludge aeration in one AR cycle, periods of aeration in the contactor and regenerator, respectively, h
t_n	necessary regeneration time or a period of one AR cycle, h
V_c,V_r	volumes of the contactor and regenerator, respectively, litre
X_c,X_m	average and initial Mixed Liquor Volatile Suspended Solids (MLVSS) concentrations in the contactor, respectively, g/l

$\bar{X}$	average concentration of MLVSS in the AR system with separate sludge regeneration, g/l
X_r	VSS concentration in recycle, g/l
p	proportion of substrate that should be oxidized to restore fully the regeneration capacity.

7.7 REFERENCES

[1] Chudoba, J., Grau, P. and Ottová, V. Control of activated sludge filamentous bulking – II Selection of micro-organisms by means of a selector. *Water Research,* 1973, 7, 1389-1406.

[2] Chudoba, J., Ottová, V. and Maděra, V. Control of activated sludge filamentous bulking – I Effect of the hydraulic regime or degree of mixing in an aeration tank. *Water Research,* 1973, 7, 1163-1182.

[3] Chudoba, J., Bláha, J. and Maděra, V. Control of activated sludge filamentous bulking – III Effect of sludge loading. *Water Research,* 1974, **8**, 231-237.

[4] Sonoda, Y., Tanaka, S. and Ishida, Y. *Journal of Fermentation Technology,* 1973, **51**, 813.

[5] Rensink, J. H. New approach to preventing bulking sludge. *Journal of the Water Pollution Control Federation,* 1974, **46**, 1888-1894.

[6] Rensink, J. H. Existence and combat of bulking sludge. *H_2O,* 1974, 7, 480–484.

[7] Heide, B. A. and Pasveer, A. Oxidation ditch: prevention and control of filamentous sludge. *H_2O,* 1974, **7**, 373-377.

[8] Rideau, J. P. and Morfaux, J. N. Decanting accidents due to bulking of activated sludge in biological conditioning of effluents of agricultural and food industries. *La Technique de l'Eau,* 1975, No. 339, 19-25, No. 340, 17-29.

[9] Morfaux, J. N., Rideau, J. P. and Serris, J. *Industries Alimentaires et Agricoles,* 1975, **92**, 4.

[10] Rensink, J. H., Jellema, K. and Ywema, T. The influence of the substrate gradient on the development of bulking sludge. *H_2O*, 1977, **10**, 14-16.

[11] Rensink, J. H. Cure and prevention of bulking sludge in practice. *Cebedeau,* 1979, **32**, 445–450.

[12] Tomlinson, E. J. and Chambers, B. *The Effect of Longitudinal Mixing on the Settleability of Activated Sludge.* Technical Report TR 122, Water Research Centre, 1979.

[13] Sezgin, M., Jenkins, D. and Parker, D. S. A unified theory of filamentous activated sludge bulking. *Journal of the Water Pollution Control Federation,* 1978, **50**, 362-381.

[14] Chapman, T. D., Matsch, L. C. and Zander, E. H. Effect of high dissolved oxygen concentrations in activated sludge systems. *Journal of the Water Pollution Control Federation,* 1976, **48**, 2486–2510.

[15] Parker, D. S. and Merrill, M. S. Oxygen and air activated sludge: another view. *Journal of the Water Pollution Control Federation,* 1976, **48**, 2511–2528.

[16] Humenick, M. J. and Ball, J. E. Kinetics of activated sludge oxygenation. *Journal of the Water Pollution Control Federation,* 1974, **46**, 735–747.

[17] Kalinske, A. A. Comparison of air and oxygen activated sludge systems. *Journal of the Water Pollution Control Federation,* 1976, **48**, 2472–2485.

[18] Boon, A. G. Oxygen activated sludge systems – I Technical Review of the use of oxygen in the treatment of waste water. *Water Pollution Control,* 1976, **75**, 206–213.

[19] Kroiss, H. and Ruider, E. Comparison of the plug-flow and completely mixed activated sludge processes. *Progress in Water Technology,* 1977, **8**, 169–174.

[20] Ulrich, A. H. and Smith, M. Operational experience with activated sludge – biosorption at Austin, Texas. *Sewage and Industrial Wastes,* 1957, **29**, 400–413.

[21] Bazyakina, N. A. *The Treatment of Concentrated Industrial Wastes,* 1958, Gosstroiizdat, Moscow. (In Russian.)

[22] Jenkins, D. and Orhon, D. in: *Advances in Water Pollution Research,* (Jenkins, S. H. ed), 1973, p. 353, Pergamon Press, Oxford.

[23] Jones, H. *Proceedings of the 5th International Conference on Water Pollution Research,* 1970, Paper No. II-5, San Francisco.

[24] Speece, R. E., Engelbrecht, R. S. and Aukamp, D. R. Cell replication and biomass in the activated sludge process. *Water Research,* 1973, **7**, 361–374.

[25] Gujer, W. and Jenkins, D. The contact stabilisation activated sludge process – oxygen utilisation, sludge production and efficiency. *Water Research,* 1975, **9**, 553–566.

[26] Dohányos, M., Grau, P. and Chudoba, J. *Proceedings of the 5th International Conference on Water Pollution Research,* 1970, Paper No. II-3, San Francisco.

[27] Yakovlev, S. V. and Karyukhina, T. A. *Vodosnab. Sanit. Tekh.,* 1966, No. 5, 36. (In Russian.)

Discussion

Chairman: **G. S. Clapham** (Wessex Water Authority, UK)

Mr. D. H. Eikelboom (Research Institute for Environmental Hygiene, TNO, The Netherlands) opening the discussion commented that Professor Grau's results indicated that the oxidation ditch should have a high accumulation capacity because of the time available for breakdown of nutrients. He thought it had been established that the accumulation capacity of an oxidation ditch was low and asked for the author's comments. In reply **Professor Grau** explained the difference between accumulation capacity and the storage capacity. Accumulation capacity of a sludge was measured by exhausting the existing substrate, adding glucose, or another polysaccharide, and calculating the capacity from the rate of uptake and the breakpoint. Accumulation capacity was relatively low, approximately 0.3 to 0.4 g/g, and depended on three factors, sludge age, proportion of saccharides in the total substrate, and type of micro-organism. There was not a great deal of data on the variation of accumulation capacity with type of micro-organism but he had found that there was definitely a difference in accumulation capacity between filamentous and non-filamentous micro-organisms. **Mr. Eikelboom** referred to Fig. 7.4 in the paper (Chapter 7) which suggested that reaeration restored accumulation capacity. However, in a system which had a long period available for reaeration, for example an oxidation ditch, the accumulation capacity of the floc was not always high. **Professor Grau** said his results showed that in a completely-mixed reactor, if more than 65 per cent of the substrate was removed, low SVI values could be achieved. These results were given in the chapter, and were for a substrate of which 50 per cent of the BOD was due to glucose. With a lower proportion of glucose in the feed promotion of the filaments would be reduced.

Professor D. Jenkins (University of California, Berkeley, USA) said that work recently completed by Dr. Lau on pure cultures of *Sphaerotilus natans* and a floc-forming organism isolated from activated sludge, had shown a relation with dissolved-oxygen concentration. The growth curves of the two types of bacteria with variation in dissolved oxygen crossed over with a K_s value for the floc former of about 0.15 mg/l of dissolved oxygen and a higher maximum growth rate than that of *Sphaerotilus natans.* Professor Jenkins also commented on the three-dimensional graph presented by the authors which plotted SVI against Dispersion Number and sludge loading rate. He understood this to show that Dispersion Number affected Sludge Volume Index (SVI) at low organic loadings but not at high organic loadings. In general he had found SVI at high loading rates to be high. He suggested that at low sludge loading rates bulking occurred due to the organisms that Mr. Eikelboom had reported in oxidation ditches, and at high sludge loadings bulking occurred because there was a deficiency in dissolved oxygen. He suggested that the bulking was due to two different sets of filamentous organisms. **Professor Grau** did not think that this was the case because the control system of the laboratory-scale plants did not allow the concentration of dissolved oxygen in the mixed liquor to fall below 2 mg/l. **Professor Jenkins** asked what the respiration rate of the sludge would be at the higher organic loadings. In reply **Professor Grau** said that respiration rates generally varied between 11 and 16 mg oxygen/g volatile mixed liquor suspended solids per hour with maximum values of up to 40 mg/g h. **Professor Jenkins** said that 2 mg of dissolved oxygen per litre would not be sufficient to prevent bulking due to low dissolved-oxygen concentrations with these respiration rates. **Dr. J. Houtmeyers** (Forest Biology Centre, Belgium) described experiments investigating intermittent and continuous feeding with different dissolved-oxygen concentrations. The experiments had shown that, generally speaking, dissolved-oxygen concentrations had no influence. When dissolved-oxygen concentration is high within the sludge floc, the effective sludge loading becomes lower for that particular sludge because of the increased proportion of active biomass. However, the most important factor was the substrate uptake rate and dissolved-oxygen concentration was of secondary importance. He had found on many occasions that bulking problems could occur in a completely-mixed system even with a high dissolved-oxygen concentration. Few problems with bulking sludge occurred with a ‘plug-flow’ system maintained at a normal dissolved-oxygen concentration. **Professor Grau** agreed that with a dense floc the concentration of dissolved oxygen in the centre could be limiting depending on the concentration of dissolved oxygen in the liquid phase and on the size and density of the floc. He agreed that this might have an effect on the sludge quality, but had found that bulking could not be suppressed by control of dissolved-oxygen concentration.

Dr. Houtmeyers asked whether Professor Grau had found any relation between substrate oxidation rates for different substrates. The work reported had been carried out with glucose but when the activated-sludge system was fed with a mixture of starch and peptone he thought there might be a difference in oxidation rate. In reply **Professor Grau** said that the rate of oxidation was related to the rate of uptake and also to the metabolic pathway of that particular substrate. It would be very different for different systems of substrate and organism and was not easy to explain in the limited time available. **Dr. Houtmeyers**

referred to the paper (Chapter 7) where it had been shown that during the uptake of glucose almost 20 per cent of the glucose was oxidized. He asked what percentage of starch was oxidized during the period of starch removal. **Professor Grau** explained in general terms the way in which starch was oxidized. The polymeric substrate was first degraded to smaller molecules such as dextrin and oligosaccharides. Some of the oligosaccharides could be transported directly into the cell, those that could not had to be broken down further to smaller molecules and finally to a substrate that could be readily transported into the cell. This resulted in a different rate of oxidation of starch and the kinetics depended upon sludge loading rate and other factors. Drawing a graph of concentration of substrate against time he showed how the concentration of polymer initially decreases as an intermediate product increases. Then the concentration of the intermediate products decreases as the concentration of readily transportable compounds, such as maltose and trisaccharides, increases. This work had been completed and it was intended to publish the results of kinetics of polymeric substrates shortly.

Mr. L. J. Webb (PIRA, UK) described work carried out by PIRA on pure cultures of *Sphaerotilus natans* in research on the growth of sewage fungus in rivers. They had used glucose and starch substrates and the results had confirmed those described by Professor Grau. Glucose was taken up immediately but with the use of starch there was a lag phase before the substrate was oxidized. **Professor Grau** said that the enzymes responsible for the breakdown of starch into readily transportable molecules seemed to be bound to the cell wall.

Mr. Webb asked Professer Grau which substrates had been used to test the theory, and whether most of the work described was based on results from glucose and oligosaccharides used as substrates. In reply **Professor Grau** described the substrates that had been used. Firstly, various monosaccharides had been tested, then a combination of two saccharides by feeding galactose to glucose adapted sludge and glucose to galactose adapted sludge. They had also tested multiple substrate combinations and finally dextrins and starch. The results from this latter work were to be published shortly.

Mr. J. H. Rensink (Agricultural University, The Netherlands) referred to work with activated-sludge plants operated at high sludge loadings (up to 5 kg BOD/kg sludge day) where low Sludge Volume Indices had been achieved. In this situation there were very short reaeration times and high accumulation capacity. He asked the author for his comments. In reply **Professor Grau** said that there were several different patterns of variation in Sludge Volume Index with loading rate. He had also found that with a completely-mixed reactor an increase in loading rate could result in the decrease in Sludge Volume Index although other authors had found the opposite. It seemed that the most successful method of suppressing bulking was by operating the process in two stages, the first stage with a high loading and the second stage with a low loading, each with its own settlement tank. In this way the first stage could be operated with a relatively low efficiency of BOD removal, and if some sludge were to be lost from the first stage a good effluent would still be produced from the second stage which would always have a good settling sludge. **Mr. Rensink** felt that his own results which had shown that a lower Sludge Volume Index could be achieved at a higher loading rate of sludge were not in agreement with the author's hypothesis. In reply **Professor Grau** did not think that Mr. Rensink's results were necessarily contradictory. He suggested that there might be another factor, maybe μ_{max} or K_s, which could play some role in a highly-loaded system. However, if a highly-loaded plant was operated continuously in a single stage mode, effluent quality would not be satisfactory. Thus, it would be better to have two stages. **Mr. Rensink** suggested an explanation for the good settleability at high sludge loadings might be due to the sludge age of the sludge being shorter than the growth rate of the filamentous micro-organisms.

Referring to the effect of phosphate on activated-sludge systems, **Mr. J. Watkins** (Yorkshire Water Authority, UK) asked for the author's comments on the effect of low phosphate concentrations in the incoming sewage on respiration rate and on uptake and adsorption. **Professor Grau** said that a relationship had been found between phosphate uptake by sludge and the Sludge Volume Index. He suggested that accumulation was probably a phosphorus-coupled reaction whereas storage would be a phosphorus-uncoupled reaction. It had been found that bacterial cells would store polysaccharides if the concentration of phosphorus or nitrogen was low. An imbalance in the nitrogen:phosphorus:BOD ratio was a well-known cause of sludge bulking and should be considered before other causes or attempts to suppress sludge bulking were looked at. **Dr. J. Houtmeyers** continued by saying that the nitrogen and phosphorus were thought to be important because when there was a deficiency the cells would store carbon. Cells containing a lot of stored carbon would take up carbon from the medium very slowly, thus affecting the accumulation rate. This phenomenon had been found in experiment on pure cultures of nitrobacter limited with nitrogen.

Dr. C. F. Forster (University of Birmingham, UK) referred to work that he had carried out sometime before which had indicated that as SVI increased the degree of polymerization of surface polymers increased. He asked whether the authors had made any analysis of the accumulated and stored polymers. In reply **Professor Grau** said that they had not carried out any experiments but data from the literature indicated that there was a difference in polymerization between stored and accumulated molecules. The stored molecules were probably highly polymerized compared with the accumulated substrate and he suspected that the accumulated substrate would be a saccharide.

Dr. E. B. Pike (Water Research Centre) referred to the differences between filamentous and unicellular micro-organisms and said that a unicell had a higher specific surface area than a filamentous organism and might be more able to scavenge for substrate. It might be expected to have a lower Michaelis constant than one growing as a filament. However, when the bacteria grow as a floc the situation changes, and limitations due to the diffusion through the floc may occur. In this case a filamentous organism which is able to grow outside the floc would now have the survival advantage because it is not limited by diffusion. **Professor Grau** said that this was a hypothesis he had considered but it had not been confirmed by experiments. Continuing the discussion **Professor H. Verachtert** (University of Louvain, Belgium) pointed out that many of the filamentous bacteria that grew out of the floc, for example *Sphaerotilus natans,* had a sheath and probably could not take up substrate except at the ends of the filaments. Thus they would not have an advantage over unicellular bacteria. This might be different with other organisms that do not have sheaths and perhaps such a distinction should be made between the different types of filaments. **Dr. A. O. Lau** (EXXON Research and Engineering, USA) had carried out experiments on *Sphaerotilus natans* and a floc-forming *Pseudomonas* isolated from activated sludge. The results had shown that *Sphaerotilus natans* had a lower K_s value than the floc-former with respect to dissolved-oxygen concentration. This would indicate that in the confines of a floc the filamentous bacteria would have an advantage over the floc-former. **Professor Grau** said that he had tried to test this hypothesis but had not been successful. The experiments had not disproved the hypothesis but had been able directly to prove it.

Mr. D. H. Eikelboom asked Dr. Lau how he had measured the K_s value of *Sphaerotilus natans* and the floc-forming bacteria and what values he had found. In reply **Dr. Lau** described the pure culture experiments on *Sphaerotilus natans* and the floc-forming organism that he had carried out. Glucose was used as the carbon substrate and the change in dissolved oxygen was monitored. A mathematical model was used to calculate K_s. The K_s value for *Sphaerotilus natans* was found to be 0.01 mg/l for dissolved oxygen and the K_s value for the floc-former was 0.15 mg/l with respect to dissolved oxygen. **Professor D. Jenkins** pointed out that the K_s values measured by Dr. Lau were for dispersed bacteria not flocs. If K_s were measured on an activated-sludge floc, the response of the organism was being measured not only to substrate concentration but also to a substrate concentration controlled by a diffusion rate. He felt that this was a mistake that was often made in activated-sludge experiments.

Chapter 8

Relation between substrate feeding pattern and development of filamentous bacteria in activated sludge

E. Van den Eynde, J. Houtmeyers and **H. Verachtert,** Catholic University of Louvain, Belgium

8.1 INTRODUCTION

The activated-sludge process has become very popular for the biological treatment of waste waters because of its flexibility and high efficiency. Filamentous bulking is a major problem in the operation of such systems. As a result of its lower compactation rate, the separation of bulking sludge during secondary sedimentation is very difficult. Sludge will not only be lost in the effluent, but it also becomes impossible to recycle enough solids to the aeration tank, resulting in an increase in sludge loading of the system. Thus the system is shifting away from it equilibrium. The increase in sludge loading will decrease the floc-forming capacity and this will lead to a further breakdown of the system.

During the last decade several authors [1, 2, 3] reported on the influence of shock loadings on sludge settleability. In 1966 it was mentioned that the settling properties of the sludge were improved by altering continuous feeding to a method which was discontinuous [4]. More details on the influence of the substrate concentration on the selection of micro-organisms in activated sludge, were given by Chudoba and co-workers [5, 6, 7]. They reported on laboratory experiments using activated-sludge systems differing in flow pattern. By introducing baffles into the aeration tank volume, different numbers of compartments were obtained. Under similar conditions of feeding and retention time the Sludge Volume Index (SVI) decreased when the substrate gradient increased together with the number of compartments in series. The high SVI in the 'completely-mixed' tank was due to an excessive growth of filamentous micro-organisms. It was concluded that 'complete-mixing', resulting in low substrate concentration throughout the aeration tank, encourages the growth of filaments in activated sludge. Aeration systems with a low degree of axial mixing, indicated by a low Dispersion Number and resulting in a higher substrate concentration gradient, were shown to promote the growth of non-filamentous micro-organisms and to suppress the growth of filamentous types. These observations were explained by supposing a different kinetic behaviour of filamentous and floc-forming micro-organisms (Fig. 8.1). In 'completely-mixed' aeration tanks, the substrate concentration is always very low (Fig. 8.1, state I). Under these conditions the growth rate of the filaments (curve A) will be higher than the growth rate of the floc-forming micro-organisms (curve B), resulting in a dominance of the filaments and leading to a higher SVI. At higher substrate concentration (Fig. 8.1, state II), as in the first compartments of the baffled system or in the first part of a 'plug-flow' reactor, the floc-forming micro-organisms will have the higher growth rate resulting in an acceptable SVI. Introducing a substrate concentration gradient over the aeration tank was advanced as a general remedy to avoid filamentous bulking. Mulder, *et al.* [1] stated that a less economical metabolism of reserve material, accumulated in the cells during the feeding phase, would explain the disappearance of filamentous bacteria in systems that receive shock loadings. This would explain the overall lower growth rate of these organisms at state II in Fig. 8.1.

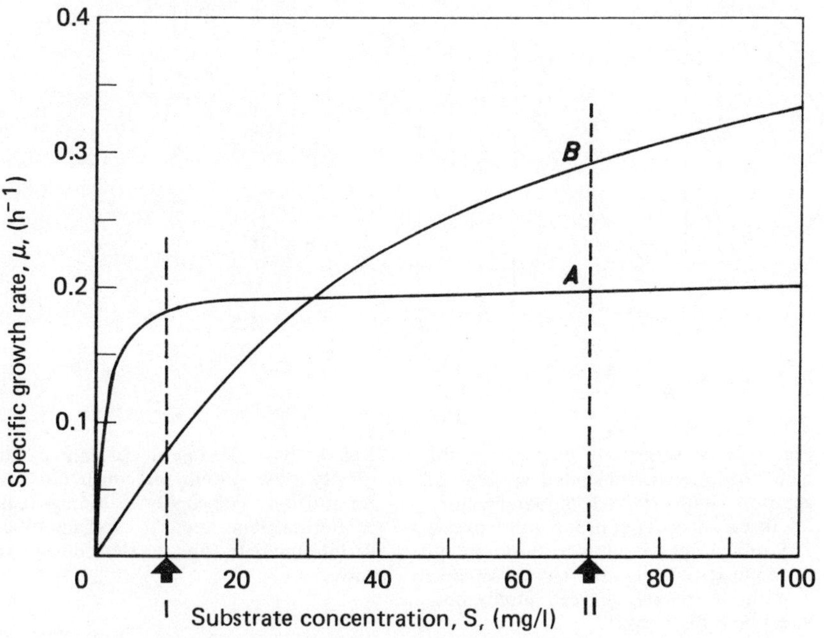

Fig. 8.1 – Grapical presentation of the relation between specific growth rate (μ) and substrate concentration (S) for two hypothetical organisms. A (K_S = 1 mg/l, $\mu_m = 0.2$ hr^{-1}) and B (K_S = 50 mg/l, $\mu_m = 0.5$ hr^{-1}) (From Chudoba *et al.* [6]).

In an attempt to obtain a better understanding of the behaviour of systems receiving shock loadings, and of the supposedly different growth kinetics of floc-forming and filamentous organisms, laboratory experiments using intermittent feeding were undertaken. Such systems allow the systematic creation of a substrate gradient in time and the existence of an endogenous phase, where the micro-organisms must rely on reserve materials. Such intermittently-fed systems were compared with normal continuously-fed systems with respect to sludge bulking, growth of filamentous bacteria and substrate removal rate. The bench-scale activated-sludge systems are represented in Fig. 8.2.

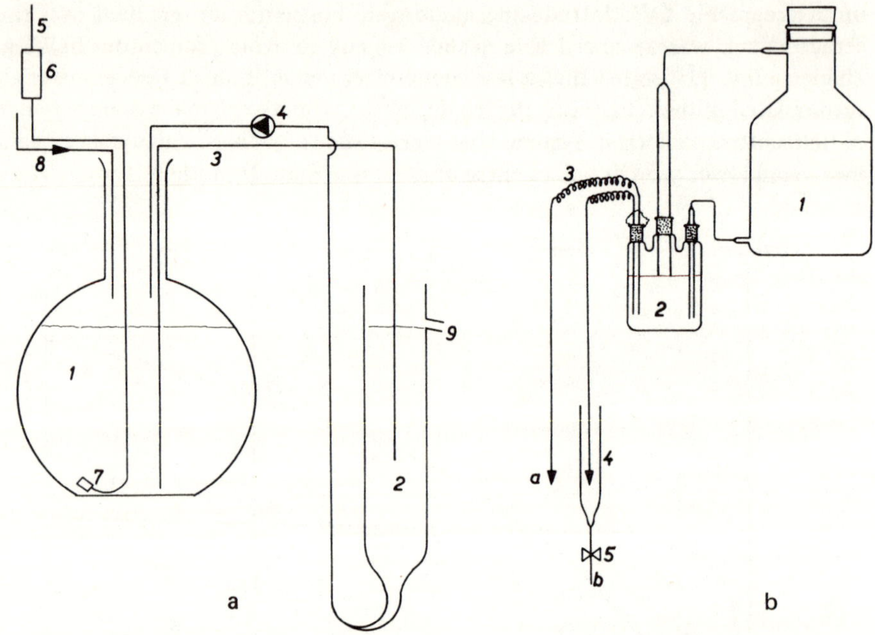

Fig. 8.2a – Schematic diagram of the bench-scale activated-sludge system. 1: aeration unit (mixed liquor volume 2.5 litres), 2: sedimentation unit (volume 175 ml), 3: siphon, 4: peristaltic pump, 5: compressed air, 6: air flowmeter, 7: ceramic sparger, 8: feed supply line, 9: effluent discharge.

Fig. 8.2b – Schematic diagram of the gravity flow system for continuous and intermittent feed supply. 1: storage tank, 2: intermediate tank, 3: capillary tubes, 4: intermediate tube, 5: electromagnetic valve.

8.2 BEHAVIOUR OF CONTINUOUSLY AND INTERMITTENTLY FED ACTIVATED SLUDGE SYSTEMS

8.2.1 Glucose and nutrient broth as organic substrates

Two types of activated sludge were fed continuously and intermittently with an artificial waste containing glucose (2.5 g/l), nutrient broth solids (0.5 g/l) and mineral salts (complete composition in Houtmeyers, *et al*, [8]). The sludge loading was adjusted to 0.4 g COD/g MLSS day and the solids retention time was about seven days. The concentration of MLSS was maintained at around 3 g/l. The variation of the MLSS and SVI with time for both sludges can be seen in Fig. 8.3. The results show that the activated sludge from the continuously-fed systems had a higher SVI than that from the intermittently-fed systems. The high SVI coincided with the presence of filamentous bacteria. Short, free floating filaments developed together with a *Sphaerotilus*-type filament radiating from the flocs. Comparing the two systems it is clear that the origin of the sludge has an influence on the selection of micro-organisms. In these experiments it was shown that the microbial composition of the activated sludge was affected by changing the method of substrate addition. These observations prompted a close investigation of the substrate profiles in the aeration vessel for different feeding patterns. Continuously-fed systems are more or less continuous cultures with a low steady state substrate concentration. To the intermittently-fed systems the substrate is added hourly, resulting in an increase in substrate concentration and initiating a substrate concentration gradient in the aeration vessel. The slope of the gradient is determined by the substrate uptake rate of the sludge. The glucose removal rate of the sludge in the activated-sludge systems described revealed that the substrate uptake rate was higher for intermittently-fed systems than for continuously-fed reactors (Table 8.1).

Table 8.1 – Glucose removal rate of intermittently and continuously-fed activated sludge

Feed pattern	Initial glucose concentration (mg/l)	Glucose removal rate (mg glucose/g MLSS min)
Intermittent	100	5.7
Intermittent	200	5.8
Intermittent	500	5.6
Continuous	200	3.4

For the intermittently-fed systems the high substrate uptake rate is responsible for the rapid exhaustion of substrate in the medium. The entire amount of substrate is taken up in less than five minutes. These results indicate the existence of two separate phases: an 'exogenous phase', during which substrates are present in solution, and an 'endogenous phase' during which there is no longer any soluble substrate available. For an activated-sludge system where two such distinct phases can be distinguished it becomes obvious that the substrate removal rate is very important for the selection of micro-organisms. It is possible that the micro-organisms with the highest substrate removal rate are able to store a larger part of the substrate during the short 'exogenous phase'. As a result growth

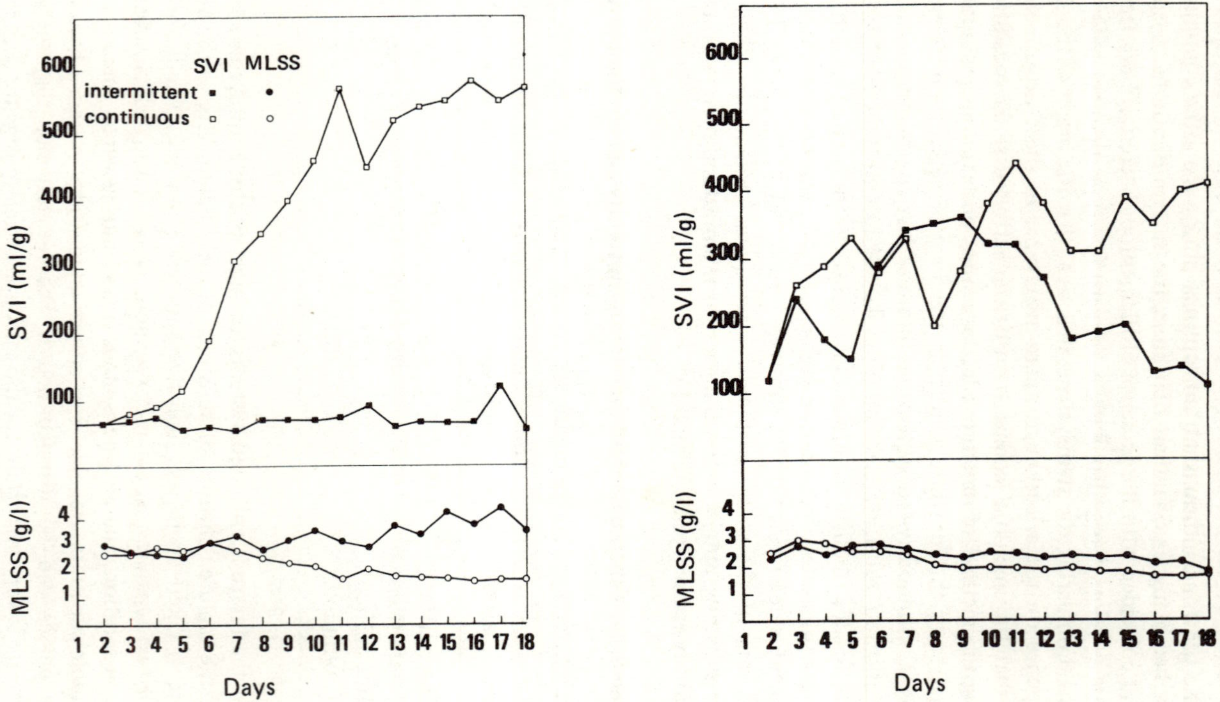

Fig. 8.3 – Characteristics of activated sludges fed intermittently and continuously with a synthetic waste containing glucose and nutrient broth and inoculated with mixed liquor from two different municipal treatment plants.

may continue during the second period when no exogenous limiting substrate is available. On the other hand, micro-organisms that are characterized by a low substrate removal rate may store only a small amount of substrate and consequently the growth conditions are not optimal during the second period when exogenous substrates are depleted. If the substrate uptake rate is smaller for filamentous micro-organisms compared to floc-forming bacteria, this could be the main reason why filamentous micro-organisms do not develop in intermittently-fed activated-sludge systems. Since this is an important factor for selection between filamentous and non-filamentous bacteria, it is interesting to compare the substrate uptake rate in parallel activated-sludge systems, fed intermittently or continuously, with different types of substrates. A comparison of substrates directly available to the cells, and substrates which must be hydrolysed by exogenous enzymes before they can enter the cells, should certainly be of interest.

8.2.2 Glucose as organic substrate

Starting with glucose as the only carbon source, both intermittently- and continuously-fed systems were operated at a sludge loading of 0.4 g COD/g MLSS day with sludge from a municipal waste water treatment plant (details in Houtmeyers, *et al.* [8]). The results of daily determinations are given in Fig. 8.4. Again the SVI of the continuously-fed system attained high values while the SVI for the intermittently-fed system remained low. The high SVI was caused by an excessive growth of filamentous micro-organisms of the *Sphaerotilus*-type. The glucose removal rate and respiration rate at the end of the experiment are given in Table 8.2.

Table 8.2 – Glucose removal rate and respiration rate of activated sludges which were fed intermittently (I) or continuously (C)

	I	C
Glucose removal rate (mg glucose/g MLSS min)	4.8	0.8
Respiration rate: (mg O_2/g MLSS min)		
endogenous	0.08	0.12
+ glucose	0.48	0.23

The higher removal rate of the intermittently-fed sludge coincides with a higher substrate respiration.

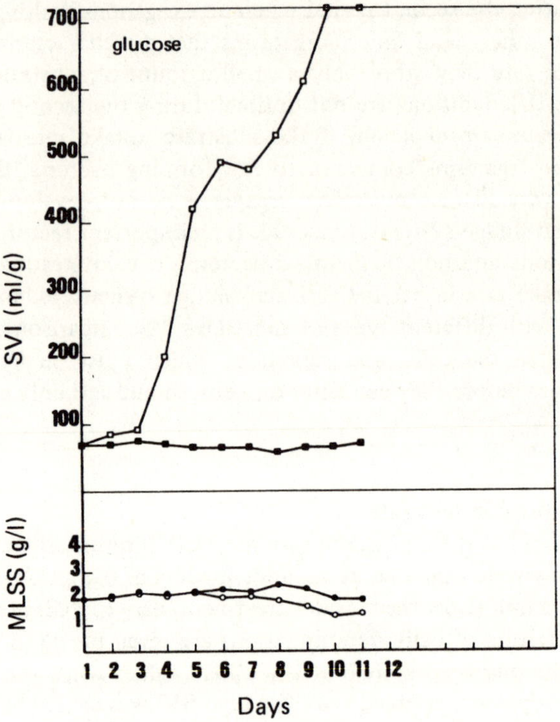

Fig. 8.4 – Characteristics of intermittently and continuously fed activated-sludge systems with glucose as the organic substrate.

8.2.3 Sodium Acetate as the only organic substrate

Two parallel activated-sludge systems, using sludge from the same origin as in the previous experiment, were maintained at the same sludge loading (Verachtert, *et al.* [9]). The variation in SVI with time is given in Fig. 8.5. *Sphaerotilus*-type filaments developed in the continuously-fed system, leading to sludge bulking. The SVI of the intermittently-fed sludge remained low.

From these experiments it can be concluded that, for substrates directly available to the cells, the concentration gradient in the aeration vessel is responsible for the selection of micro-organisms in activated sludge. Another series of experiments was performed with carbon sources of the polymer-type; starch and casein were used. Again bench-scale systems were run comparing continuous feeding with intermittent feeding.

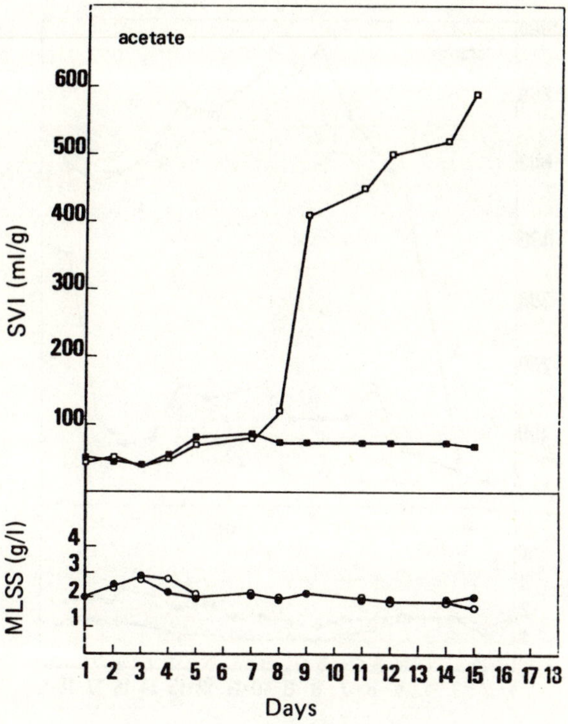

Fig. 8.5 – Characteristics of intermittently and continuously fed activated-sludge systems with sodium acetate as the organic substrate.

8.2.4 Starch as the only organic substrate

For starch, the sludge settleability in the two systems is given in Fig. 8.6. The SVI value for the continuously-fed system increased rapidly and after five days of operation very high values were achieved; again filamentous bulking was the cause. The SVI of the sludge from the intermittently-fed system had variable values and tended to increase towards the end of the experiment; after 15 days of operation a considerable amount of filaments developed in the system and after 20 days of operation the intermittently-fed system contained almost as many filaments as the continuously-fed system. The results on the substrate removal and respiration rates of both sludges are given in Table 8.3.

The starch removal rate for the sludge from the intermittently-fed system is surprisingly high; however, the respiration rate of the sludge after starch addition is only twice the endogenous respiration rate. In previous experiments the substrate respiration rate of the intermittently-fed system was normally several times higher than the endogenous respiration rate. These results may indicate that the

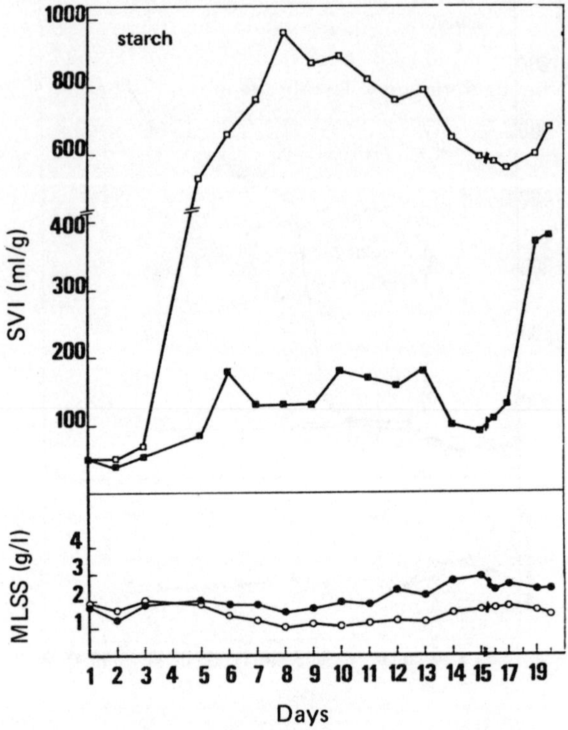

Fig. 8.6 – Characteristics of intermittently and continuously fed activated-sludge systems with starch as the organic substrate.

Table 8.3 – Substrate removal rate and respiration rate of activated sludges fed intermittently (I) or continously (C) with starch

	I	C
Substrate removal rate (mg substrate/g MLSS min)		
starch	3.2	0.4
glucose	2.6	0.9
Respiration rate (mg O_2/g MLSS min)		
endogenous	0.10	0.13
+ starch	0.19	0.28
+ glucose	0.29	0.23

starch was removed rapidly from solution but that it was adsorbed to the flocs. After hydrolyses of the adsorbed starch, the end products could then enter the cells. If this is what occurs then it is possible that the hydrolysis of the starch is the rate limiting step and the concentration gradient of the substrate that can be assimilated by the bacteria may largely be absent. The starch concentration gradient in the aeration vessel would lead to a gradient in glucose concentration which is much less pronounced and selection of micro-organisms may not occur.

8.2.5 Casein as the only organic substrate

In a subsequent experiment casein was used. The results of daily analyses are given in Fig. 8.7. It is shown that the SVI is very low in both systems. During this experiment filamentous micro-organisms did not develop and deflocculation occurred in both systems. It is interesting to note that deflocculation of dairy waste activated sludge was also mentioned by Pipes [10]. Further investigation revealed that there was no physical action of casein on the flocs. Deflocculation could be the result of high protease activity of sludge fed with casein.

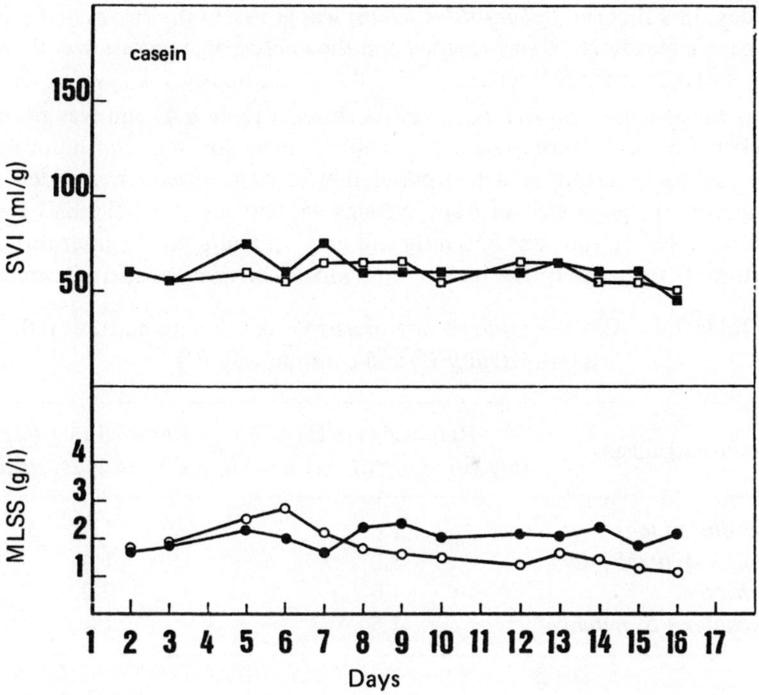

Fig. 8.7 – Characteristics of intermittently and continuously fed activated-sludge systems with casein as the organic substrate.

As a general conclusion it can be stated that with directly absorbable substrates, a striking correlation exists between the feed pattern, the type of micro-organisms growing in the activated sludge, and the substrate removal rate. For polymeric substrates the rate of enzymatic hydrolysis may be the determining parameter. All these results are in good agreement with our hypothesis. However, the supposition that the substrate removal rate is higher for floc-forming than for filamentous micro-organisms has not been proved unequivocally until now. Pure culture experiments were initiated to collect more data.

8.3 INTERMITTENTLY AND CONTINUOUSLY-FED PURE CULTURES

Sphaerotilus natans was chosen as a typical filamentous bacterium and an *Arthrobacter* strain as a typical floc-forming micro-organism. Both strains had been isolated from activated sludge. In two chemostats, one adapted for continuous feeding and one for intermittent feeding, pure and mixed cultures of *Arthrobacter* and *Sphaerotilus* were compared. Glucose was always the only carbon source and vitamins were added as growth factors for *Sphaerotilus*, when an inorganic nitrogen source was used. The dilution rate of the systems was always low (0.0167 hr^{-1}), corresponding to a sludge loading of 0.4 g glucose/g MLSS day. In a first run *Sphaerotilus natans* was grown in the two systems. After a few days a steady state was reached and the microscopic picture was the same for the intermittently-fed system as for the continuously-fed system. In both cultures the glucose removal rate was measured (Table 8.4) and was found to be higher for the intermittently-fed culture than for the continuously-fed culture. To each culture an active population of *Arthrobacter* was added. This *Arthrobacter* sp. grew well in both systems. A few days later the SVI for the continuously-fed culture was 850 ml/g and only 150 ml/g for the intermittently-fed culture. The glucose uptake rate for both mixed cultures was also determined.

Table 8.4 – Glucose removal rate of various continuous cultures fed intermittently (I) and continuously (C)

micro-organisms	Removal rate (I) (mg glucose/g dry wt min)	Removal rate (C) (mg glucose/g dry wt min)
Sphaerotilus natans	1.8	1.2
S. natans + Arthrobacter	3.2	1.4
Arthrobacter	4.6	2.2
Arthrobacter + S. natans	4.5	1.4

In a second run *Arthrobacter* sp. was grown in the two systems. After a few days floc-formation was observed and the glucose uptake rate was again higher in the intermittent system (Table 8.4). After one week of continuous culture *Sphaerotilus natans* was added to the cultures. In the intermittently-fed system *Sphaerotilus* did not grow, while in the continuously-fed culture this filamentous micro-organism caused bulking, leading to a higher SVI. The glucose uptake rates for all the systems can be compared in Table 8.4.

The results of the pure and mixed culture studies revealed that the substrate uptake rate of *Sphaerotilus natans* is always lower than the substrate uptake rate of *Arthrobacter* sp.; which clearly confirms our hypothesis. The data in Table 8.4 show that the way in which the substrate is added to the culture influences the glucose uptake rate. It was repeatedly found that pure cultures of micro-organisms have a higher substrate removal rate with intermittent feeding than with continuous feeding. This means that the substrate uptake rate is not constant for a specific micro-organism, but can be manipulated by the feeding pattern or by the growth conditions in general. Induction of glucose uptake mechanisms by the higher glucose concentration or by the alternation of high and low substrate concentrations in the intermittently fed cultures may be possible.

8.4 DISCUSSION

The influence of the feeding pattern on the species compostion of activated sludge can be explained in terms of substrate uptake rate. Under regimes with a period of high substrate concentration, floc-forming micro-organisms might develop more rapidly than filamentous micro-organisms, because of their higher substrate removal rate (Fig. 8.8, state II). The growth of filamentous bacteria in continuously-fed systems can be understood by supposing a higher substrate uptake rate for filaments at low substrate concentrations (Fig. 8.8, state I). However, on examining a sample of bulking sludge, one can always find a high proportion of floc-forming micro-organisms. In continuously-fed systems, filaments never completely displace the floc-forming species. This means that both types can grow equally at low substrate concentrations. For *Arthrobacter* and *Sphaerotilus* the saturation constants, K_m, for glucose uptake were determined on cells growing in continuous cultures at low dilution rate ($0.04 hr^{-1}$). For *Arthrobacter*, values were recorded between 1.5 and 5 mg/l and for *Sphaerotilus* between 1 and 3 mg/l. The different values for K_m in different runs are an indication that K_m might not really be a constant. It was shown that the two micro-organisms have almost the same affinity for the substrate, which means that in continuously-fed cultures a mixed population of filamentous and floc-forming bacteria may grow.

We also reported that the change from continuous feeding to intermittent feeding of a pure culture caused an increase in the substrate uptake rate. This might indicate that adaptation or induction mechanisms are active. If changing to an intermittent feed results in a stronger induction of permeases in floc-forming micro-organisms than in filamentous micro-organisms, the uptake rate of the former will be higher in intermittently-fed systems. This mechanism would also explain how the creation of a concentration gradient in the aeration volume of an activated-sludge system can cure filamentous bulking.

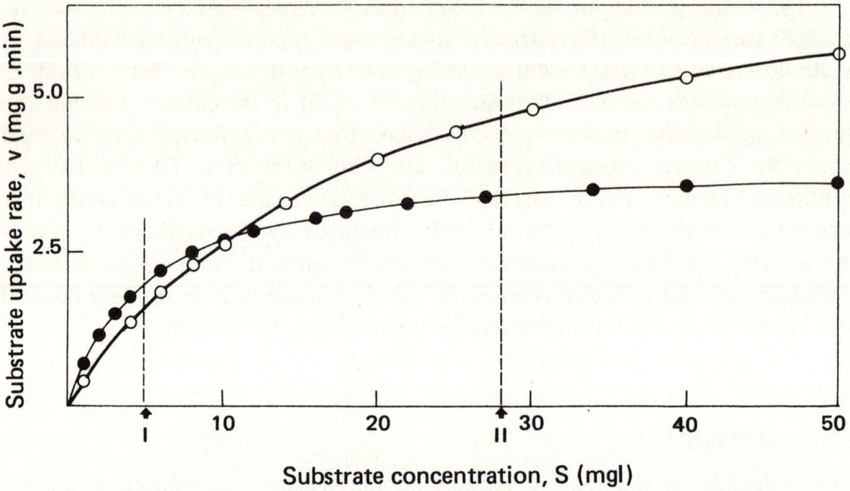

Fig. 8.8 – Substrate uptake rate (v) as a function of the substrate concentration (S) for two hypothetical organisms. A floc-former (v_m = 8 mg glucose/g MLSS min; K_m = 20 mg/l) and a filamentous bacterium (v_m = 4 mg glucose/g MLSS min; K_m = 5 mg/l).

It is not clear how the sequential operation of intermittent systems affects the physiology of the bacteria. In the exogenous phase, with a relatively high substrate concentration in the medium, the substrates are taken up very rapidly by the micro-organisms. Probably reserve materials are then formed on which the bacteria can grow and survive in the endogenous phase. It is known that different micro-organisms can produce different types of reserve materials. Adamse [11] found that *Arthrobacter* could accumulate large amounts of polysaccharides; up to 50 per cent of the cell dry weight. Mulder and Van Veen [12] found that *Sphaerotilus* could synthesize poly-β-hydroxybutyrate (29 per cent), polysaccharides (38 per cent) and lipids (4 per cent). Micro-organisms surviving on different reserve materials will show different decay kinetics. It was shown that *Sphaerotilus natans* died off faster than *Arthrobacter* under the same conditions [11]. Mulder and co-workers [1], concluded that both floc-forming and filament-forming micro-organisms could produce reserve materials in the exogenous phase, but that the filamentous bacteria were outgrown by the floc-forming bacteria because of a less economical metabolism of their reserve materials. For another filamentous micro-organism, *Haliscomenobacter hydrossis,* Krul [13] found that this bacterium could not produce reserve substances, even in the presence of a high concentration of glucose. For this particular micro-organism it was believed that glucose had a repressive effect on the uptake of amino acids, thus limiting its growth. In our investigations only carbon compounds were considered and further research is necessary on the influence of nitrogen compounds. Nothing is known about the uptake rate of nitrogen compounds in an intermittently-fed system. If the high uptake rate of the carbon source causes a relative deficiency in nitrogen for the cell, metabolism could be switched towards the production of reserve materials. Subsequently, the species will compete for the available nitrogen compounds. Further research on the influence of feeding pattern on the physiology of the micro-organisms remains necessary.

8.5 REFERENCES

[1] Mulder, E. G., Antheunisse, J. and Crombach, W. H. Microbial aspects of pollution in the food and dairy industries, in: *Microbial Aspects of Pollution, Society of Applied Bacteriology Symposium Series No. 1* (eds. Sykes, G. and Skinner, F. A.), 1971, pp. 71–89, Academic Press, London.

[2] Pasveer, A. An investigation on development and control of filamentous sludge. *H_2O*, 1969, **2**, 28–34.

[3] Rensink, J. H. New approach to preventing bulking sludge. *Journal of the Water Pollution Control Federation,* 1974, **46**, 1888–1894.

[4] R. A. A. D. *Verslag Proef Oxydatiesloten Over de Periode 1963–1966,* 1966, Rijkszuivel – Agrarische Afvalwaterdienst, Arnhem.

[5] Chudoba, J., Ottová, V. and Madĕra, V. Control of activated sludge filamentous bulking – I Effect of the hydraulic regime or degree of mixing in an aeration tank. *Water Research,* 1973, **7**, 1163–1182.

[6] Chudoba, J., Grau, P. and Ottová, V. Control of activated sludge filamentous bulking – II Selection of micro-organisms by means of a selector. *Water Research,* 1973, **7**, 1389–1406.

[7] Chudoba, J., Bláha, J. and Madĕra, V. Control of activated sludge filamentous bulking – III Effect of sludge loading. *Water Research,* 1974, **8**, 231–237.

[8] Houtmeyers, J., Van den Eynde, E., Poffe, R. and Verachtert, H. Relation between substrate feeding pattern and development of filamentous bacteria in activated sludge processes – I Influence of process parameters. *European Journal of Applied Microbiology and Biotechnology,* 1980, **9**, 63–77.

[9] Verachtert, H., Van den Eynde, E., Poffe, R. and Houtmeyers, J. Relation between substrate feeding patterm and development of filamentous bacteria in activated sludge processes – II Influence of substrate present in the influent. *European Journal of Applied Microbiology and Biotechnology,* 1980, **9**, 137–149.

[10] Pipes, W. O. Microbiology of dairy waste activated sludge separation problems. *Proceedings of 32nd Industrial Waste Conference, Purdue University.* 1977, 303–309. Science Publishers Incorporated, Ann Arbor.
[11] Adamse, A. D. Bulking of dairy waste activated sludge. *Water Research,* 1968, 2, 715–722.
[12] Mulder, E. G. and Van Veen, W. L. Investigations on the *Sphaerotilus Leptothrix* group. *Antonie van Leeuwenhoek,* 1963, **29**, 121.
[13] Krul, J. M. Experiments with *Haliscomenobacter hydrossis* in continuous culture without and with *Zoogloea ramigera. Water Research,* 1977, **11**, 197–204.

Discussion

Chairman: **G. S. Clapham** (Wessex Water Authority, UK)

Mr. G. F. G. Clough (Howard Humphreys and Partners, Leatherhead, Surrey, UK) opened the discussion by asking Mr. Van den Eynde at what temperature the experiments were carried out. **Mr. Van den Eynde** said that the activated-sludge experiments were carried out at room temperature, which varied in the range 18 to 21 °C and the pure culture experiments were carried out in the first instance at room temperature, with determination of K_m values performed at 29 °C in a chemostat.

Mr. J. L. Hutton (Grampian Regional Council, Aberdeen, UK) said that he was interested in the intermittent feeding mentioned by Mr. Van den Eynde. He asked how frequently the feeding was carried out and what percentage of the daily dose was given at any one time. He asked how this compared with the feeding regime of a full-scale treatment plant dealing with a diurnal variation in flow. Mr. Hutton also asked if Mr. Van den Eynde could confirm that sludge was wasted from the experimental plant and that the same sludge loading rate was maintained throughout. He also made the comment that it was common practical experience that when a plant was started up there was often a period of several months before filamentous bacteria appeared and bulking occurred. He therefore wondered whether the intermittent feeding experiments which were carried out over such a short period as 20 days would fully represent what might have happened over a longer period. He commented finally that Mr. Van den Eynde and his colleagues had selected *Sphaerotilus natans* as a typical filamentous bacterium, but earlier papers had shown that it only occurred in about 10 per cent of bulking problems and he therefore wondered whether they would be repeating the work with other filamentous bacteria such as *Haliscomenobacter hydrossis* or Type 1701.

Mr. Van den Eynde replied that they intended to repeat some of their experiments using other filamentous bacteria. He said that the continuously-fed cultures were fed in a drop wise manner but in the intermittently fed plant the drops were collected over a one hour period and added as one dose at the end of the hour. He added that the continuously-fed system always failed after 20 days. His colleague, **Dr. J. Houtmeyers**, said that in some of the experiments the plants were operated for 40 to 50 days. Intermittently-fed plants did not bulk during this period but the continuously-fed plants were bulking after 20 days of operation.

Professor D. Jenkins (University of California, Berkeley, California, USA) asked what the growth rate was during the experiments. **Mr. Van den Eynde** replied that the sludge age in the experiments was seven days. **Professor Jenkins** said that in his experience it took three sludge ages for steady-state to be reached following plant start-up, and this being the case some of the experiments might not have been run for long enough. **Mr. Van den Eynde** said that in some of the experiments filamentous bacteria started to develop after only two to three days.

Mr. E. I. Clark (North West Water Authority, UK) asked how the authors had maintained the same MLSS concentration in the two systems. He also asked Mr. Van den Eynde if he could indicate how their findings could be applied to full-scale plant operation. **Mr. Van den Eynde** replied that they had maintained the same MLSS concentration in the two systems by measuring the MLSS concentration each day and calculating how much sludge to waste. In the case of the continuously-fed plant, which was bulking, they had allowed for the sludge which was being wasted in the effluent. He said that for full-scale applications it was not practical to apply their technique of intermittent feeding. He would recommend the use of a plug-flow contact mixing tank positioned before the aeration units. He said that they had further research to perform in order to determine the lowest substrate gradient necessary for the correct selection of micro-organisms. Following this research the dimensions of the contact tank could be specified more accurately.

Dr. J. Houtmeyers said that the intermittently-fed system was really a laboratory-scale simulation of a 'plug-flow' system. He said that they had preferred to use intermittent feeding to simulate plug-flow because when they had used a series of mixed compartments problems with wall-growth had arisen and different bacteria had developed in the compartments. He said that a contact tank with a retention time of ten minutes was adequate.

Dr. A. O. Lau (EXXON Research and Engineering, New Jersey, USA) asked Mr. Van den Eynde whether the dissolved oxygen concentration in the reactors was measured during the pilot-plant studies. He also asked Mr. Van den Eynde how he had assessed when steady-state conditions had been achieved in the pure culture experiments. Dr. Lau had carried out similar experiments to measure K_m values and he had found that he needed to run the experiments for three to four weeks before steady-state was achieved and he could measure K_m. Mr. Van den Eynde had said that he measured the K_m values after a few days. Dr. Lau queried whether this was a long enough period to achieve steady-state. Dr. Lau closed by commenting that the K_m values he had found were

10 mg/l for the floc-forming bacteria and 14 mg/l for *Sphaerotilus natans* both measured at 20 °C.

Mr. Van den Eynde replied that they had not measured the dissolved oxygen concentration in the chemostats continuously, but they were operated with long retention times and with low solids concentrations and hence they were sure that the dissolved oxygen concentration was quite high. He said that they had found that it usually took about three retention times to reach steady-state conditions and he did not feel it was necessary to wait for three or four weeks, since after that time bacteria started to grow on the walls and rotor blades of the chemostat. He did not think that the system would then be under steady-state conditions because of the heavy wall growth.

Mr. A. G. Boon (Water Research Centre) said that he wished to follow-up the earlier questions by Messrs Hutton and Clark relating to the use of the data collected by Mr. Van den Eynde in operating full-scale plants. He said that normally there would be variations in the substrate concentrations entering a plant, from hour to hour in large plants and even from minute to minute in a small plant. He asked if Mr. Van den Eynde would expect a plant which received a wide diurnal variation in rate of flow and loading to have a better settling sludge than a plant with a balanced flow rate and sewage composition.

Also, he asked, would Mr. Van den Eynde expect a small, rural, activated-sludge plant, which receives a wide variation in flow, to have a better settling sludge than a large regional works where the diurnal variation in flow will have been smoothed out.

Mr. Van den Eynde said that when a completely-mixed system was used and the concentration changes in the influent sewage were large, there would be high gradients throughout the system and the effluent quality would be poor. He said that to overcome this an exogenous period was needed in a contact tank followed by an endogenous period in the aeration tank.

Mr. H. Slijkhuis (Agricultural University, Wageningen, The Netherlands) referred to Table 8.3 and asked why the respiration rate of starch in the intermittently-fed systems was only twice the endogenous rate. Mr. Van den Eynde had said that it was likely that the enzymes used for degrading starch were rate-limiting but in the continuously-fed systems this would not be the case. He also asked if Mr. Van den Eynde could explain why the substrate-removal rate for glucose in the intermittently-fed system was 2.6 mg/g MLSS min in Table 8.3, while in Table 8.1 it was 5.7 mg/g MLSS min. Mr Slijkhuis asked if he was to conclude that Mr. Van den Eynde had measured the substrate removal rate of glucose in Table 8.3 at concentrations lower than 30 mg/l because he had said that even at 30 mg/l the glucose removal rate was 5.7 mg/g MLSS min, the maximum rate.

Mr. Van den Eynde, replying to the second question, said that the reason for the difference was that the systems described in Table 8.3 were adapted to starch and not to glucose. In Table 8.1 the systems were adapted to glucose.

In reply to Mr. Slijkhuis's first question, Mr. Van den Eynde said that he did not think that there was a significant difference between 0.19 mgO_2/g MLSS min, the respiration rate of the intermittently-fed plant using starch and 0.28 mgO_2/g MLSS min, the rate for the continuously-fed plant. He said that the values were dependent upon the activity of the sludge but it was also possible that minor errors in dry-weight determination could have affected the lower value.

Chapter 9

The influence of feed pattern on sludge bulking

J. H. Rensink, H. J. G. W. Donker and **T. S. J. Ijwema,** Agricultural University of Wageningen, The Netherlands

9.1 INTRODUCTION

The activated-sludge process has been used widely to purify waste water to a high effluent quality. The activated-sludge flocs must obey two quite distinct conditions. They must first remove the organic material from the water. In addition, the activated-sludge solids should be separated from the purified water in the settling tank. The first requirement – removal of organic material – presents few problems, in general. However, there are often problems with the separation of the activated-sludge solids from the treated effluent. In many activated-sludge plants the bulking sludge phenomenon is encountered. Bulking sludge has poor settling characteristics and compactability and gives rise to a high Sludge Volume Index (SVI), (that is the volume in millilitres occupied by one gram of activated sludge mixed liquor solids, dry weight, after settling for 30 minutes in a 1000 ml graduated cylinder).

Bulking sludge has been the subject of much research. According to Metcalf and Eddy [1] the causes that are most commonly cited in the literature are related to:

- the physical and chemical characteristics of the waste water,
- treatment plant design limitations, and
- plant operation.

Waste water characteristics that can affect sludge bulking include fluctuations in flow and strength, pH-value, temperature, staleness, nutrient content, and the nature of the waste components. Design limitations include air-supply capacity, clarifier design, return-sludge pumping capacity limitations, and short circuiting or poor mixing. Operational causes include low dissolved oxygen in the aeration tank, organic waste overloading of the aeration tank and final clarifier operation.

Special attention has been given to the filamentous bacteria in activated sludge by Van Veen [2], Van Veen *et al.* [3] and Eikelboom [4].

A new hypothesis about the cause of bulking sludge has been developed by Rensink [5, 6], and Chudoba *et al.* [7, 8, 9]. Rensink *et al.* found that bulking sludge from oxidation ditches treating waste water could be cured in the laboratory. The bulking sludge improved when it was fed in a batch system. These results suggested that the feed pattern was more important than the sludge loading. In a preliminary experiment a batch system was compared with a 'completely-mixed' system treating synthetic dairy waste water. The sludge loading employed was 300 g BOD_5/kg MLSS day and the Mixed Liquor Suspended Solids (MLSS) concentration varied between 2 to 2.5 g/l. The settleability of the sludge in both systems was measured after 5, 15 and 30 minutes. The results are depicted in Fig. 9.1. It is clear that the sludge settleability in the 'completely-mixed' system began to deteriorate after about one week. The filamentous bacteria *Sphaerotilus natans* prevailed in the activated-sludge floc. However, the sludge of the batch system was compact, stable and contained hardly any filamentous bacteria.

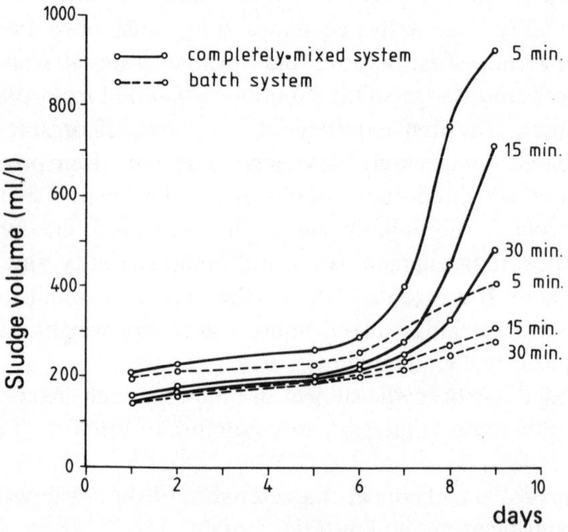

Fig. 9.1 – The variation of settleability of activated sludge measured after 5, 15 and 30 minutes in a batch system, and a 'completely-mixed' system at a sludge loading of 300 g BOD_5/kg MLSS day.

This result gave rise to experiments to investigate the relationship between feed pattern and bulking sludge more extensively. This chapter gives a summary of our research on this topic and deals with experiments conducted under laboratory and pilot-plant conditions.

9.2 LABORATORY STUDIES

9.2.1 Conventional activated-sludge plants

The first experiments were performed on

(i) a batch system,
(ii) a 'completely-mixed' system, and
(iii) a 'plug-flow' system.

All the systems were fed with synthetic sewage at loadings in the range 25 to 2000 g BOD_5/kg MLSS day. The batch system consisted of an aeration compartment with a volume of 12 litres (Fig. 9.2). For the 'completely-mixed' system an aeration tank and sedimentation tank with capacities of 3 and 2 litres respectively were used. The 'plug-flow' system consisted of a long and narrow aeration tank, divided in 12 compartments of equal volume, with a total capacity of 6 litres. 'Plug-flow' conditions could be achieved in this aeration tank. From dye-tracer response curves it was determined that the aeration tank was equivalent to 10 ideally-mixed compartments in series. The sedimentation tank had a volume of 2 litres (Fig. 9.2). The activated sludge used in these experiments was taken from a batch system fed with synthetic waste water. During the experiments the MLSS was maintained at 2 g/l. Excess sludge was wasted every day, and the dissolved oxygen was never limiting. The composition of the synthetic sewage is given in Table 9.1.

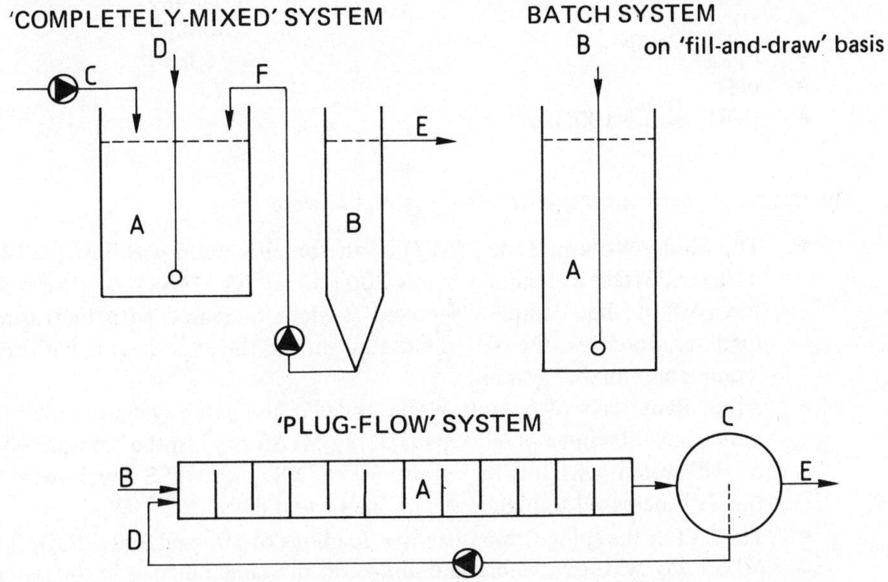

Fig. 9.2 – The laboratory units. Batch system: A – aeration tank, B – air supply and diffuser stone. 'Completely-mixed' system: A – 3-litre aeration tank, B – 2-litre sedimentation tank, C – feed line, D – air supply and diffuser stone, E – effluent, F – return sludge. 'Plug-flow' system: A – 6-litre aeration tank, B – feed line, C – 2-litre sedimentation tank, D – return sludge, E – effluent.

Table 9.1 – Composition of synthetic sewage

Toilet-paper	4 mg/l
Sunlight soap	3 mg/l
Urea	30 mg/l
Gelatine	60 mg/l
Starch	300 mg/l
Skimmed milk	180 mg/l
$NaHCO_3$	150 mg/l
Na_2HPO_4	10 mg/l
$MgSO_4$ $7H_2O$	3 mg/l
$FeCl_3$ 6 H_2O	4 mg/l
KCl	4 mg/l

All the components were dissolved in tap water with a total hardness of 70 mg $CaCO_3$/1. The synthetic waste water revealed the following characteristics:

- COD 556 mg/l
- BOD_5 350 mg/l
- $N_{Kjeldahl}$ 35 mg/l
- P_{total} 6 mg/l
- pH 7.6
- BOD_5:N:P = 100:10:2

The results of these experiments – see Fig. 9.3 – were:

- The Sludge Volume Index (SVI) of the batch system was low, (< 100 ml/g) and stable at loadings below 300 g BOD_5/kg MLSS day. However, the SVI of the 'completely-mixed' system increased with increasing loading. Moreover, the SVI fluctuated during the experiments with the 'completely-mixed' system.
- After many days of operation the sludge in the batch system tended to bulk above loadings of 300 g BOD_5/kg MLSS day. In the 'completely-mixed' system with loadings above 300 g BOD_5/kg MLSS day, however, the SVI increased to high levels in one or two days.
- The SVI in the 'plug-flow' system at loadings of 100 and 300 g BOD_5/kg MLSS day was low, stable and almost in the same range as in the batch system.
- The filamentous micro-organisms in the bulking sludge were *Sphaerotilus natans, Flexibacter* or *Haliscomenobacter.* Usually one of these micro-organisms was predominant.

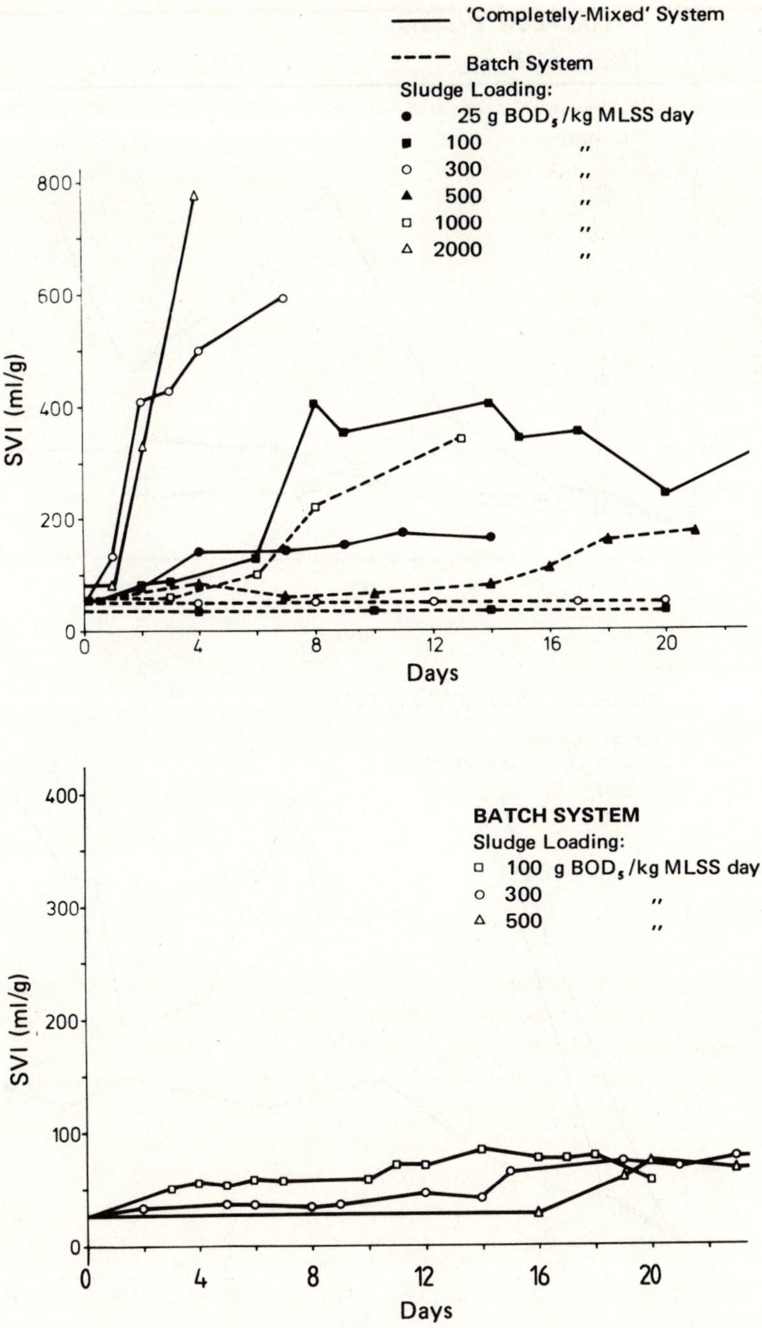

Fig. 9.3 – The variation of the SVI at different loadings (*continued over page*).

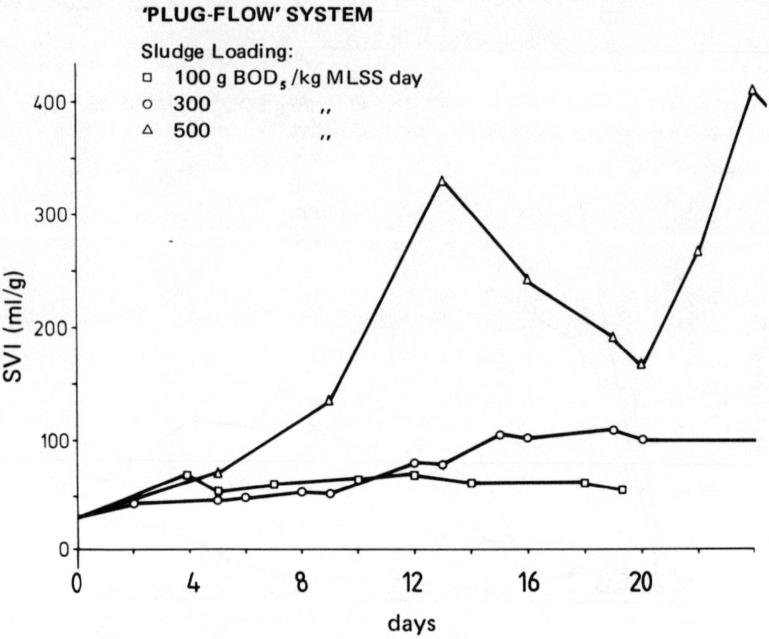

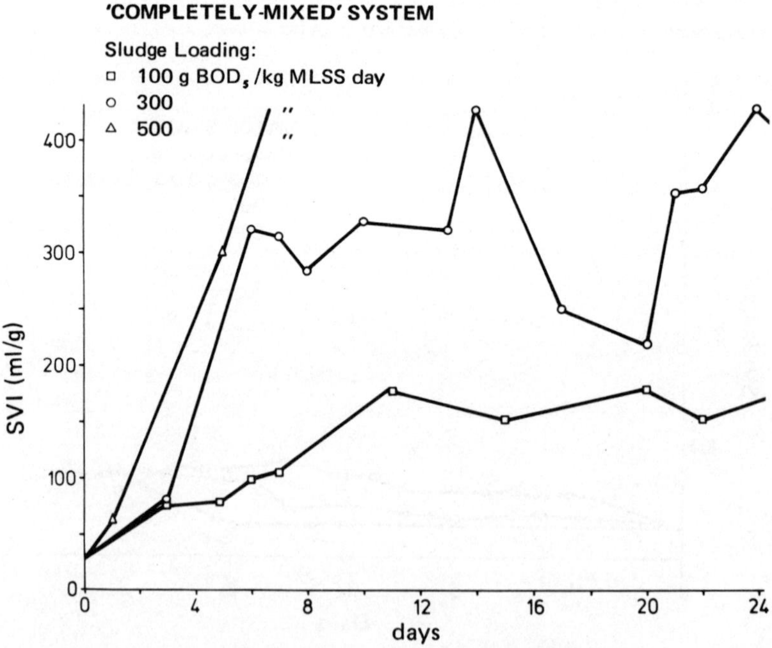

Fig. 9.3 (*continued*) – The variation of the SVI at different loadings.

9.2.2 Modification of conventional activated-sludge plants

Laboratory experiments were carried out to study the influence of modifications of conventional activated-sludge plants on bulking sludge. All the plants were operated at loadings of 300 g BOD_5/kg MLSS day. The following process modifications were employed:

(i) continuous 'completely-mixed' system;
(ii) continuous 'plug-flow' system;
(iii) batch system: 'plug-flow';
(iv) semi-continuous system: step loading;
(v) semi-continuous system: step sludge.

For the 'completely-mixed' and the 'plug-flow' systems the same equipment as described in the previous section was used.

To simulate (iii) an ideal 'plug-flow' system, (iv) a step-loading system and (v) a step-sludge system, the laboratory units were operated as follows:

(iii) Ideal 'plug-flow'

Ideal 'plug-flow' conditions were simulated by operating the plant as a fill-and-draw system.

(iv) Step loading

A modification of a conventional plant is step loading. In a conventional plant the waste water enters at the beginning of the aeration tank, together with the return-sludge. In the step-loading system the waste water is introduced lengthwise along the aeration tank at different points. Thus the sludge becomes more and more diluted from the inlet of the aeration tank to the outlet. At every point in the aeration tank the activated sludge receives more waste water. At the outlet the purified effluent and activated sludge are separated. This system can be simulated as follows:

> A tank filled with aerated activated sludge is fed continuously with waste water. Thus, the sludge becomes more and more dilute in a similar manner to an aeration tank with step loading. After a given time aeration and feeding are stopped, the sludge settles and the effluent is drawn off the tank. This can be called a semi-continuous system since the feeding is continuous and the effluent discharge is a batch process. Thus step loading is simulated by a semi-continuous system with continuous feeding of waste water.

(v) Step sludge

This process is characterized by the introduction of the return-sludge length wise along the aeration tank at various points. To simulate this system semi-continuously the feed is added batch wise and the sludge is added continuously. In this semi-continuous system the sludge is stored in another aerated tank. A small part of the activated sludge remains in the original aerated tank. All the waste water is added to this activated sludge. After two hours the stored activated sludge is pumped into the aerated tank. Thus step sludge is simulated by a semi-continuous system with continuous feeding of 'return-sludge'.

For these experiments the synthetic waste water consisted of glucose (360 mg/l) and the same inorganic salts as mentioned in Table 9.1. The aerated compartment for these semi-continuous systems had a capacity of 12 litres and a MLSS concentration of 2 g/l. Excess sludge was wasted daily. The initial sludge was obtained from a batch reactor fed with the synthetic waste water.

The results of these experiments – see Fig. 9.4 –were:

- The batch system (ideal 'plug-flow') had an SVI which decreased during the experiment to 50 ml/g. The sludge floc was stable and compact.
- The semi-continuous system with step loading induced bulking sludge after eight days of operation. *Sphaerotilus natans* prevailed.
- The semi-continuous system with step sludge gave a stable and low SVI. No filamentous bacteria occurred.
- The continuously fed 'completely-mixed' system produced a bulking sludge. *Sphaerotilus natans* was predominant.
- The continuously fed 'plug-flow' system had a stable and compact sludge floc.

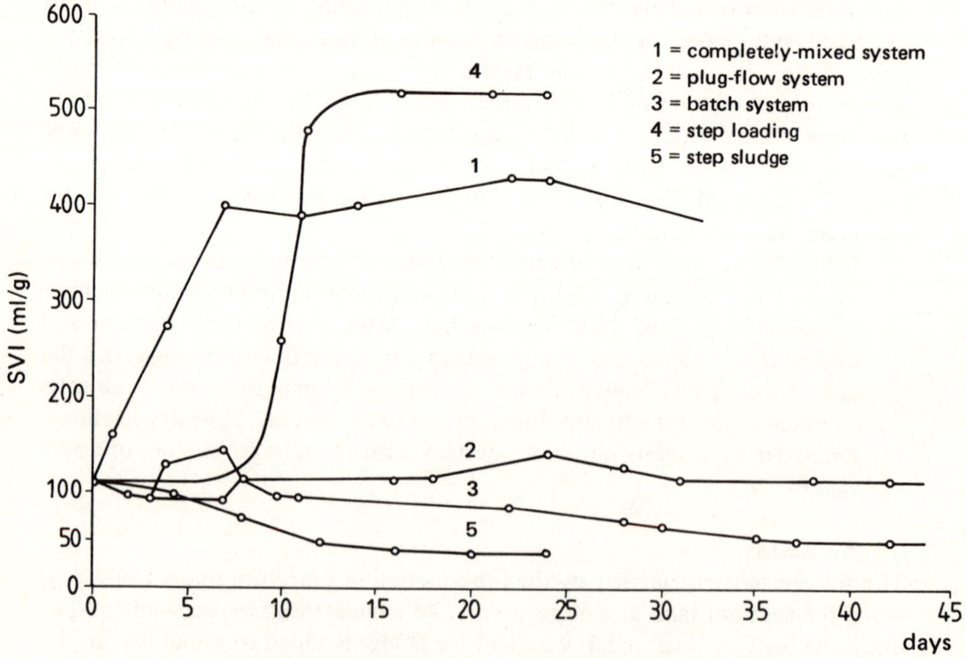

Fig. 9.4 – The variation of SVI, at a loading of 300 g BOD_5/kg MLSS day, in different activated-sludge systems.

9.3 PILOT PLANT STUDIES

9.3.1 'Completely-mixed' and 'plug-flow' systems

The occurrence of bulking activated sludge has been studied extensively in pilot-plants [13]. Two separate systems at different loadings were used (Fig. 9.5):

(i) a 'completely-mixed' aeration tank with a capacity of 1000 litres;
(ii) a 'plug-flow' aeration tank with a capacity of 1000 litres.

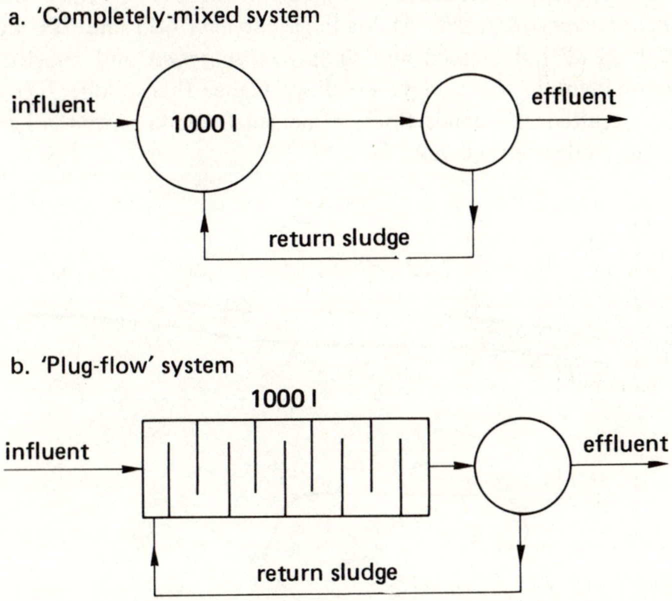

Fig. 9.5 – Pilot-plant units a. 'Completely-mixed' system b. 'Plug-flow' system.

The units were fed with settled domestic sewage. At dry weather flow the following average compostion was measured.

- COD 500 mg/l
- BOD_5 300 mg/l
- $N_{Kjeldahl}$ 70 mg/l
- P_{total} 18 mg/l
- Alkalinity 7 meq/l
- Ca 50 mg/l
- Mg 2.5 mg/l
- Hardness 125 mg $CaCl_3$/l
- BOD_5:N:P = 100:23:6

During the experiments the MLSS concentration was maintained at 2 g/l. Excess sludge was wasted continuously and the dissolved oxygen was never limiting.

In Fig. 9.6 the SVI is given as a function of the loading for the 'completely-mixed' system (CM) and the 'plug-flow' system (PF). At loadings in the range 100 to 600 g BOD_5/kg MLSS day the CM-system induced bulking sludge. In the filamentous sludge *Sphaerotilus natans* prevailed. Below a loading of 100 g BOD_5/kg MLSS day the SVI was mostly low. In some cases bulking sludge occurred and the filamentous micro-organisms responsible were *Microthrix parvicella* or *Haliscomenobacter hydrossis*. At loadings between 600 and 700 kg BOD_5/kg MLSS day the SVI decreased and *Sphaerotilus natans* and zoogloeal bacteria were predominant in the sludge. Loadings higher than about 700 kg BOD/kg MLSS day resulted in smaller flocs, dispersed bacteria and increasingly turbid effluent. The sludge settled well.

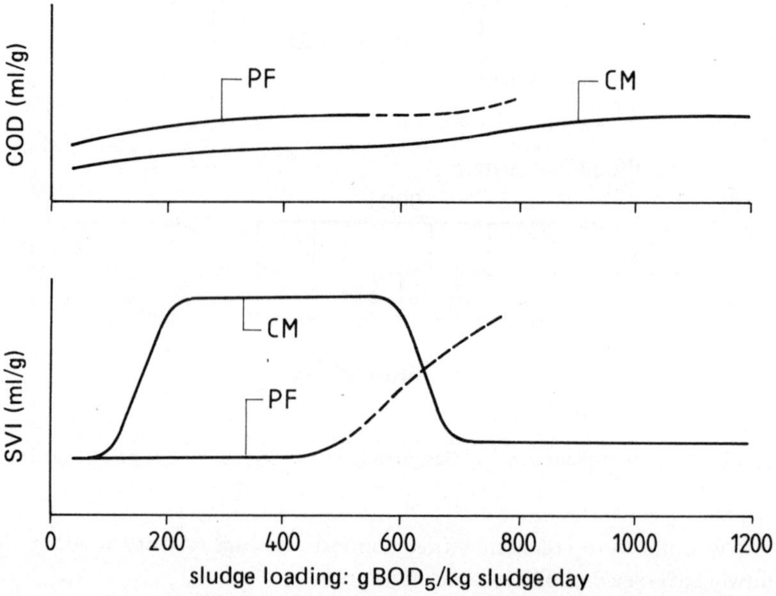

Fig. 9.6 – Relationship between SVI and sludge loading in the 'completely-mixed' system (CM) and the 'plug-flow' system. Relationship between the COD of the filtered mixed liquor in CM and first compartment PF at different loadings.

The 'plug-flow' system had a low and stable SVI below a loading of 500 g BOD_5/kg MLSS day. The floc was compact and few or no filamentous micro-organisms were present. Slightly above a loading of 500 g BOD_5/kg MLSS day the sludge became bulky and *Sphaerotilus natans* and zoogloeal bacteria occurred massively. Higher loadings have not been investigated.

The COD of the filtered mixed liquor from the CM-system and from the first compartment of the PF-system is presented in Fig. 9.6. The soluble COD in the first compartment of the 'plug-flow' system was higher than the soluble COD in the 'completely-mixed' system. The difference in COD could be the main factor responsible for the occurrence of filamentous micro-organisms. This will be discussed subsequently.

9.3.2 'Complete-mixing' with contact tank

Heide and Pasveer [14] and Rensink [10] have drawn attention to the fact that the SVI can be improved if the waste water and the return-sludge passes through a small contact tank before entering a 'completely-mixed' aeration tank. In the contact tank return-sludge and waste water are mixed and the sludge is heavily loaded with organic material. Heide and Pasveer have introduced the concept of 'loading level', which is defined as the mass of COD, or BOD, entering the contact tank per hour divided by the mass of return-sludge passing the contact tank per hour (g COD/kg MLSS or g BOD_5/kg MLSS).

In practice, on large scale plants, a small contact tank ahead of the treatment plant does not guarantee an improvement of the SVI. Therefore experiments were performed with different types of contact tank and with different waste waters, namely a synthetic dairy waste water and settled domestic sewage.

(a) Synthetic dairy waste water

A rectangular tank and a 'completely-mixed' tank were tested. Each was placed ahead of an oxidation ditch with a capacity of 240 litres. The contact tanks each had a volume of 12 litres. See Fig. 9.7.

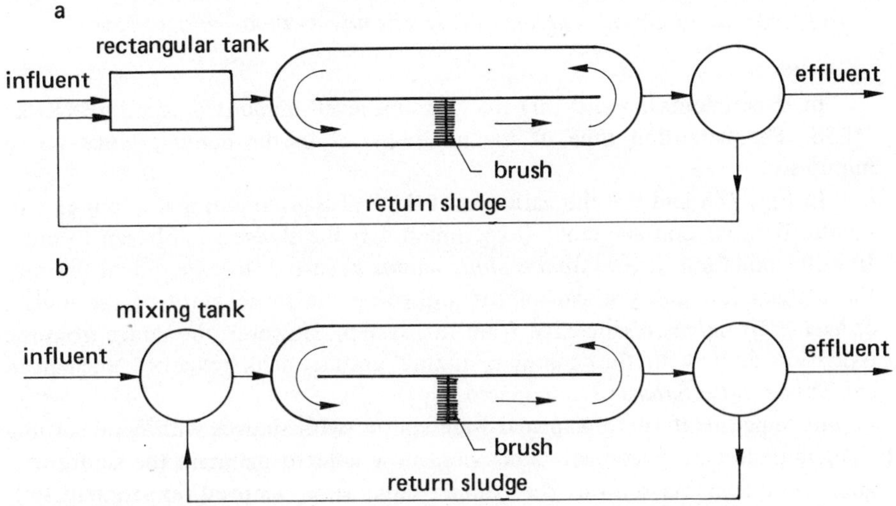

Fig. 9.7 – Pilot-plant units a. Oxidation ditch with a rectangular contact tank b. Oxidation ditch with a 'completely-mixed' contact tank.

Every day the pilot plants were supplied for ten hours with synthetic dairy waste water. This waste water was prepared by mixing whey powder and skimmed milk powder (ratio 3:1) and dissolving in tap water. The synthetic waste water had the following characteristics:

- COD 1250 mg/l
- BOD_5 750 mg/l
- $N_{Kjeldahl}$ 35 mg/l
- P_{total} 7 mg/l
- pH 7.6
- BOD_5:N:P = 100:5:1

Each pilot plant was fed with 80 litres of waste water for ten hours only, each day. The loading amounted to 120 g BOD_5/kg MLSS day. The MLSS concentration was maintained at 2 g/l, excess sludge was removed daily, and the dissolved oxygen was never the limiting factor. The following experiments were performed:

(i) *Operation of the pilot-plants without contact tanks.* The experiment was started with a sludge with good settling properties, adapted to the synthetic dairy waste water.

(ii) *Operation of the oxidation ditches with the contact tanks.* The experiment was started with bulking sludge, the result of experiment (i).

(iii) *Operation of the plants with contact tanks.* The experiment was started with a sludge with good settling properties, adapted to synthetic dairy waste water.

(iv) *Continuation of experiment (iii), but now without contact tanks.*

In experiments (i) and (iii) the 'loading level' amounted to 60 g COD/kg MLSS. The detention time of the mixed-liquor in the contact tanks was 8 minutes.

In Figs. 9.8 and 9.9 the variation of the SVI is given during the four experiments. Without contact tanks (experiment (i)) the sludge soon began to bulk. In both oxidation ditches *Sphaerotilus natans* occurred. In experiment (ii) only the rectangular tank was capable of improving the settleability of the sludge. *Sphaerotilus natans* disappeared from this system. However, the sludge from the oxidation ditch with the 'complete-mixing' contact tank remained unchanged and *Sphaerotilus natans* predominated.

In experiment (iii) the plants were started with sludges with good settling characteristics. Both contact tanks were now able to maintain the sludge in a good condition. As soon as the contact tanks were removed (experiment (iv)) the sludge in the oxidation ditches bulked due to massive growth of *Sphaerotilus natans.*

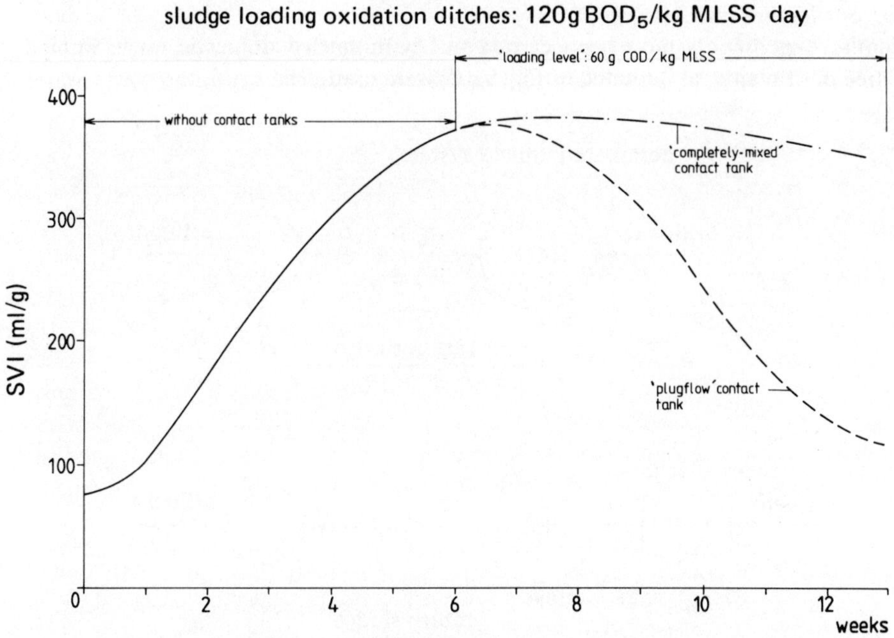

Fig. 9.8 – The variation of the SVI with and without contact tanks.

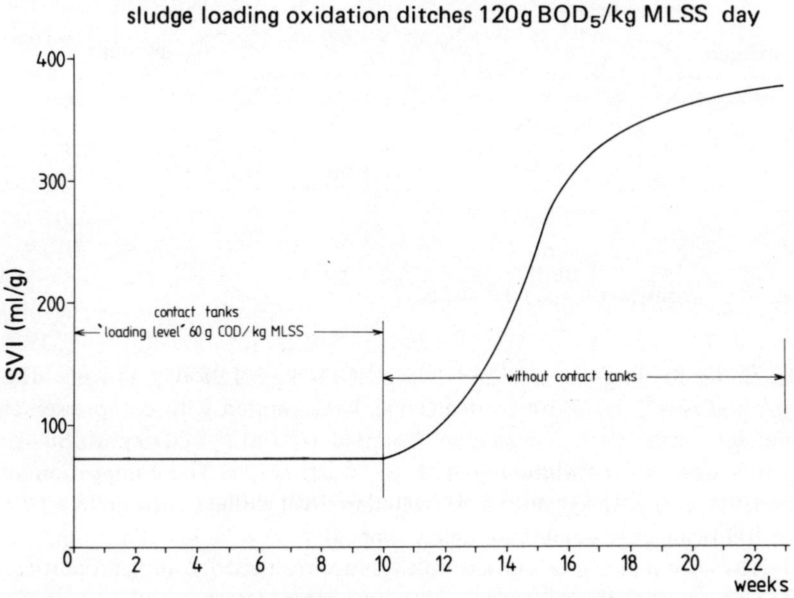

Fig. 9.9 – The variation of the SVI with and without contact tanks.

(b) Settled domestic sewage

Similar experiments have been carried out with settled domestic waste water. Three pilot plants, as depicted in Fig. 9.10, were used.

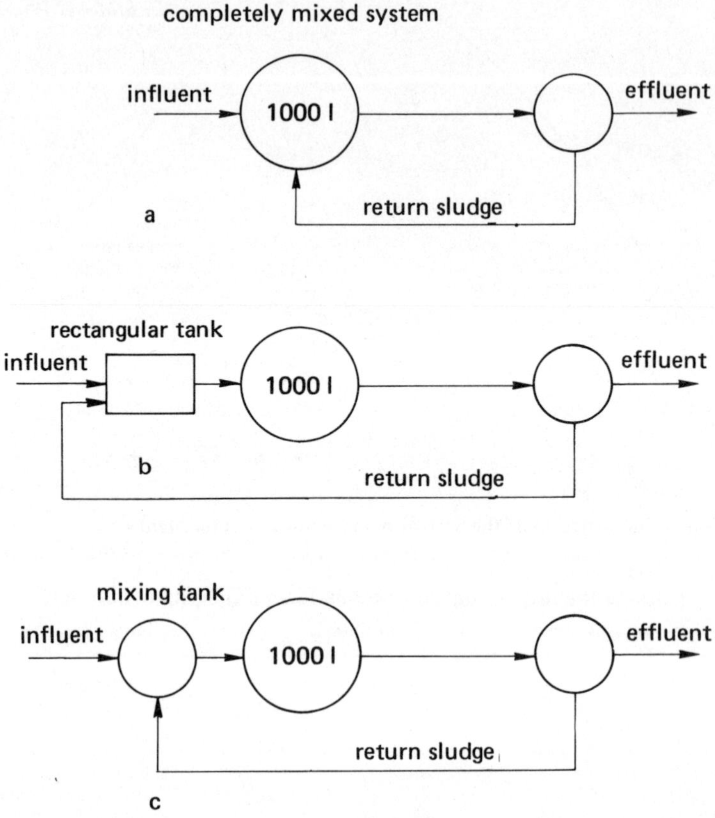

Fig. 9.10 – Configuration of pilot plants.

The sludge loading amounted to 225 g BOD_5/kg MLSS day and the MLSS concentration was 2 g/l. Both contact tanks were aerated with compressed air. With contact tanks the loading level amounted to 120 g COD/kg MLSS. The pilot plants were fed continuously with the waste water. The compostion and characteristics of the waste water have been described earlier.

The following experiments were performed:

(i) The pilot plants were started with a sludge with good settling properties.
(ii) Operation of the pilot plants without contact tanks.
(iii) Operation of the pilot plants with contact tanks.

In Fig. 9.11 the variation in the SVI is given during the three experiments. In experiment (i) only the pilot plant without the contact tank induced bulking sludge. Both contact tanks were capable of preventing bulking sludge. In experiment (ii), without contact tanks, both pilot plants produced bulking sludge. In experiment (iii) only the rectangular tank had the capacity to cure the bulking sludge.

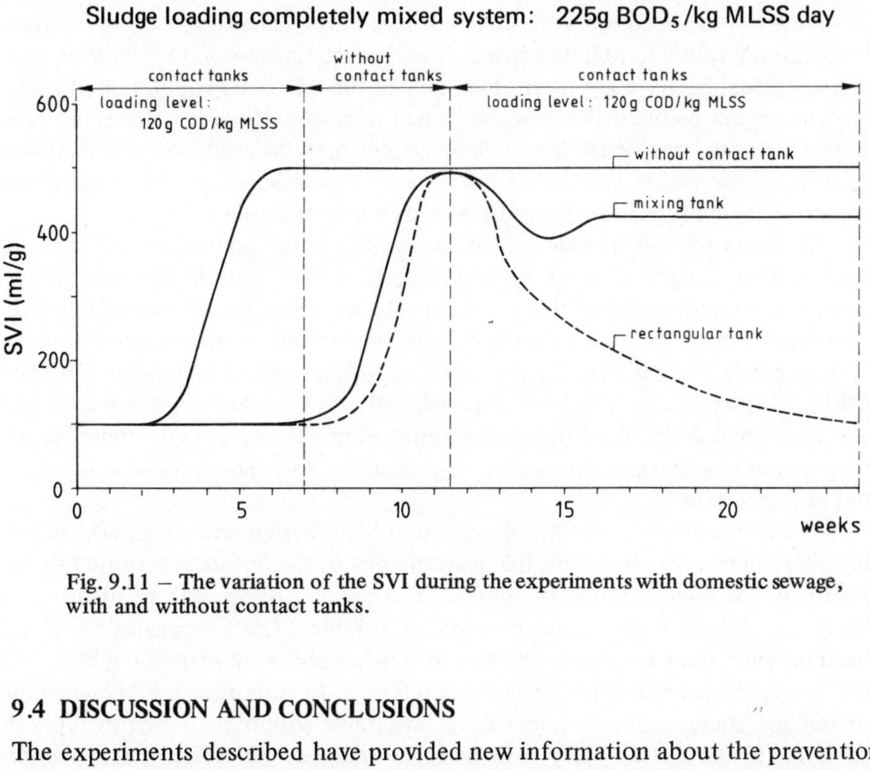

Fig. 9.11 – The variation of the SVI during the experiments with domestic sewage, with and without contact tanks.

9.4 DISCUSSION AND CONCLUSIONS

The experiments described have provided new information about the prevention of bulking sludge. It has been proved that the method of waste water supply to an activated-sludge plant plays a decisive role in the prevention of bulking sludge. Bulking occurred in systems where the sludge was only in contact with a low concentration of soluble COD. This was the case in the 'completely-mixed' systems (laboratory and pilot-plant) and the semi-continuous system with step loading. Sludges with good settling properties were produced by systems where the sludge was in temporary contact with a high concentration of soluble COD. This situation existed in the 'plug-flow' continuously fed systems (laboratory and pilot-plants), in batch systems, in the semi-continuous system with step sludge and in the contact tanks ahead of the oxidation ditches.

The 'loading level' (g COD/kg MLSS) as defined by Heide and Pasveer [14] can be related to the conditions at the inlet of the 'plug-flow' system and the conditions in step sludge and batch systems. In an activated-sludge plant started with a sludge which settles well, the type of contact tank does not matter. Both types investigated prevented the formation of bulking sludge. However, the configuration of the contact tank is important in curing bulking sludge. In this case a contact tank with a rectangular configuration is necessary. Pilot-plant studies with settled domestic waste water gave the same results. This investigation has not yet been completed.

How can the occurrence of bulking sludge be explained? *Sphaerotilus natans* is commonly found in polluted rivers. According to Curtis *et al.* [15] *Sphaerotilus* and zoogloeal bacteria are often dominant in the microbial population of slime-growths in the polluted rivers of the United Kingdom. However, slimes found in waters with an organic carbon content greater than 20 mg/l and/or a dissolved oxygen concentration less than 5 mg/l, were frequently dominated by zoogloeal bacteria. These bacteria are non-filamentous micro-organisms.

In slimes formed in waters with an organic carbon content of less than 20 mg/l and an oxygen concentration of greater than 5 mg/l *Sphaerotilus natans* usually prevailed. *Sphaerotilus* – and probably the other filamentous micro-organisms too – are able to grow at low concentrations of organic nutrients. This is partly due to their large surface to volume ratio. Liebmann [16] and Curtis [15] stated that *Sphaerotilus* will only grow in oxygen-rich waters and was not found at dissolved oxygen concentrations below 2.5 mg/l. Under laboratory conditions *Sphaerotilus* is able to grow at 0.1 to 2.0 mg/l dissolved oxygen, but at a slow rate.

The above information on the growth of *Sphaerotilus* and zoogloeal bacteria in rivers can be compared with the microbiology of the different activated-sludge plants. A low concentration of soluble COD induces the growth of filamentous bacteria, while a higher concentration of soluble COD propagates zoogloeal bacteria with good settling properties. At loadings of more than 300 g BOD_5/kg MLSS day the soluble COD concentration is high. In such cases the phenomenon of bulking sludge can be explained by favourable conditions for both types of bacteria. If the COD is very high in a continuously fed system, *Sphaerotilus natans* will be replaced by zoogloeal bacteria due to the very short sludge age. This situation also occurs in a 'completely-mixed' system, with a continuous feed.

This study has shown that the main factor causing bulking sludge is the feed pattern. In a 'plug-flow' system the activated sludge has, in general, a lower SVI than in a 'completely-mixed' system. Therefore use of 'plug-flow' systems is recommended.

9.5 REFERENCES

[1] Metcalf and Eddy Inc. *Wastewater Engineering: Treatment, Disposal, Reuse,* 1979, McGraw-Hill, New York.

[2] Van Veen, W. L. The bacteriology of activated sludge, in particular the filamentous bacteria. *Antonie Van Leeuwenhoek,* 1973, **39**, 189.

[3] Van Veen, W. L., Van der Kooy, D., Geuze, E. C. W. A. and Van der Vlies, A. W. Investigations on the sheathed bacterium *Haliscomenbacter hydrossis* isolated from activated sludge. *Antonie Van Leeuwenhoek,* 1973, **39**, 207.

[4] Eikelboom, D. Filamentous organisms in activated sludge. *Water Research,* 1975, **9**, 365-388.

[5] Rensink, J. H. Verslag Proefoxydatiesloten Over de Periode 1965-1966, *Rijks Agrarishce Afvalwaterdienst,* 1966, Arnhem, The Netherlands.

[6] Rensink, J. H. Le chlore naissant pour la lutte contre *Sphaerotilus natans* dans la boue activée. *Revue Laitière Francaise,* 1966, **234**, 444–449.

[7] Chudoba, J., Ottová, V. and Maděra, V. Control of activated sludge filamentous bulking – I Effect of the hydraulic regime or degree of mixing in an aeration tank. *Water Research,* 1973, **7**, 1163-1182.

[8] Chudoba, J., Grau, P. and Ottová, V. Control of activated sludge filamentous bulking – II Selection of micro-organisms by means of a selector. *Water Research,* 1973, **7**, 1389-1406.

[9] Chudoba, J., Bláha, J. and Maděra, V. Control of activated sludge filamentous bulking – III Effect of sludge loading. *Water Research,* 1974, **8**, 231-237.

[10] Rensink, J. H., Voetberg, J. W. and Ijwema, T. S. J. Die Verhinderung der Eintwicklung von Blahschlamm bei der Vollbiologischen Reinigung von Abwasser der Pommesfritesindustrie. *GWF-Wasser/Abwasser,* 1977, **118**, 75-79.

[11] Rensink, J. H. New approach to preventing bulking sludge. *Journal of the Water Pollution Control Federation,* 1974, **46**, 1888-1894.

[12] Rensink, J. H. De Invloed van het Voedingspatroon op het Ontstaan van Licht Slib bij Verschillende Slibbelastingen. *H_2O,* 1974, **7**, 480–484.

[13] Rensink, J. H. Cure and Prevention of bulking sludge in practice. *La Tribune du Cebedeau,* 1979, **432**, 445–450.

[14] Heide, B. A. and Pasveer, A. Oxidation ditch: prevention and control of filamentous sludge. *H_2O,* 1974, **7**, 373-376.

[15] Curtis, E. J. C. and Curds, C. R. Sewage fungus in rivers in the United Kingdom: the slime community and its constituent organisms. *Water Research,* 1971, **5**, 1147-1159.

[16] Liebmann, H. *Handbuch der Frischwasser und Abwasserbiologie,* 1965, R. Oldenbourg, Munich.

Discussion

Chairman: **G. S. Clapham** (Wessex Water Authority, UK)

The discussion was opened with a question from **Mr. J. J. Preedy** (Hawker Siddeley Water Engineering Limited, UK) who asked Mr. Rensink about the physical characteristics, size, length-to-width ratio and 'plug-flow' characteristics, of the rectangular contact tank used for his pilot-scale experiments. **Mr. Rensink** replied that the rectangular contact tank had a volume of 12 litres and a length of 1.0 m. The area of cross-section was 120 cm^2. A perforated pipe was installed in the bottom of the contact tank to provide aeration. No measurement of the plug-flow characteristics had been made, but a substrate concentration gradient was measured during the experiments.

Mr. L. Rennerfelt (Ferriklor International AB, Sweden) asked Mr. Rensink to explain why the results using the rectangular contact tank were so much better than those using the circular, completely-mixed tank. Mr. Rensink replied that both tanks had similar retention times, but that the rectangular tank exhibited a low degree of dispersion. Therefore the latter provided a high substrate concentration at its inlet and a low concentration at its outlet. Although he had no definite evidence the author thought that this substrate gradient, from inlet to outlet, must be responsible for a higher loading of the sludge at the inlet of the contact tank and therefore a higher degree of adsorption, which stimulates the growth of bacteria with good settling properties.

Mr. L. Barnes (Anglian Water Authority, UK) said that we had now heard quite a lot of information about the advantages of contact tanks and premixing tanks upstream of activated-sludge aeration tanks, and he confirmed that this had been his own experience also. However, he had obtained better results when such tanks were aerated, with either coarse-bubble or fine-bubble aeration. He wondered whether the authors had performed any tests to quantify the differences between the performance of aerated and non-aerated contact tanks or premixing tanks. **Mr. Rensink** replied that these tests had been performed and that a 10 per cent improvement in effluent quality had been obtained with aerated tanks. However, sludge settleability did not vary significantly. He reported also that slightly better effluent quality was achieved with rectangular contact tanks and this was probably a result of the improved sludge settleability. **Mr. D. H. Eikelboom** (Research Institute for Environmental Hygiene, TNO, The Netherlands) said that he thought the improved effluent quality achieved by the rectangular contact tanks could be explained by the high rates of adsorption which can be achieved in these tanks as a result of their 'plug-flow' characteristics.

Mr. I. W. Devenish (Southern Water Authority, UK) pointed out that in the work described in the paper (Chapter 9) the influent and recycled sludge were added to the contact tank as separate streams and he wondered whether further benefits might be obtained if the streams were combined before entering the contact tank. **Mr. Rensink** replied that this possibility had not been investigated. **Mr. Devenish** then asked for clarification of the values of loading given in Fig. 9.8. He was told that the figure of 120 g BOD_5/kg MLSS d related to the overall sludge loading of the oxidation ditch whilst the value of 60 g COD/kg MLSS represented the 'loading level' in the contact tank, the latter term having been defined in the paper.

Professor P. Grau (Prague Institute of Chemical Technology, Czechoslovakia) presented a diagram showing that an aeration tank consisting of eight separate compartments could be equivalent to between 1.5 and 9.8 theoretical ideal mixing tanks, depending on how the tanks are arranged and the size of connections between them. He stressed therefore that it was very important to perform stimulus-response tests with tracer substances in order to establish the dispersion characteristics of contact tanks, premixing tanks and entire aeration tanks.

In reply to a question from **Mr. G. F. G. Clough** (Consultant, Howard Humphries and Partners, UK) **Mr. Rensink** said that Fig. 9.6 did not indicate that 'plug-flow' plants produced effluents of lower quality than completely-mixed systems. The line marked PF referred to COD values in the first compartment of a 'plug-flow' plant and the values in the final compartment were very similar to those for the completely mixed system, indicated as CM.

The final point in the discussion was raised by **Mr. L. J. Webb** (PIRA, UK) who referred to the authors' comparison of the development of bulking sludge with the formation of sewage fungus in rivers. He pointed out that from his own work in this area measurements such as BOD, COD and TOC are not satisfactory indicators of the growth of organisms such as *Sphaerotilus natans* in rivers. Of much greater importance is the specific nature of the organic material present, particularly the molecular weight which affects the method of assimilation.

Chapter 10

Effect of longitudinal mixing and anoxic zones on settleability of activated sludge

B. Chambers, Water Research Centre, Stevenage Laboratory

10.1 INTRODUCTION

The first chapter of this book has indicated that poor settleability occurs continuously or intermittently in about half of the activated-sludge plants in England and Wales [1]. There have been many attempts to relate the settleability of activated sludge to the operating conditions of the plant or the characteristics of the waste water being treated. These relationships generally refer to individual plants and may not be applicable to other plants. This chapter describes two aspects of process design which can be employed widely to improve the settling characteristics of activated sludge.

10.1.1 The anoxic mixing stage

Anoxic mixing stages were originally introduced into activated-sludge plants to effect denitrification. The simplest form of plant which can be employed to bring about partial denitrification is a conventional activated-sludge plant with an anoxic stage at the inlet to the aeration tanks. Within this stage, settled sewage and recycled sludge are mixed, usually by a low-speed mechanical stirrer. It has been observed that the sludges produced in plants with anoxic stages usually settle well [2, 3]. Pilot-scale experiments to investigate the use of an anoxic stage to control the settleability of activated sludge have been performed by the Water Research Centre (WRC) and published as a Technical Report [4]. This chapter contains a summary of the results obtained.

10.1.2 Longitudinal mixing characteristics

The selection of flow patterns within chemical or biological reaction vessels is normally made in accordance with the kinetics of the reactions involved. The extreme complexity of the kinetics of reactions involved in waste water treatment has, in the past, resulted in a somewhat irrational choice of flow patterns within activated-sludge aeration tanks. Although the terms 'plug-flow' and 'complete-mixing' are used frequently to describe these flow patterns, the terms actually describe ideal situations which are unlikely to occur in full-scale practice. In all real situations flow patterns or mixing characteristics are more correctly described by reference to the degree of backmixing, longitudinal mixing or dispersion.

Recently there has been a great deal of interest in the effect of aeration tank mixing characteristics on sludge settleability. Chudoba and co-workers [5] were the first to attempt to quantify the mixing characteristics of activated-sludge aeration tanks and to relate the values obtained to the settling characteristics of the sludges produced. In order to express the degree of longitudinal mixing exhibited by pilot-scale plants they used a Dispersion Number as described by Levenspiel [6]. For the ideal cases of 'plug-flow' and 'complete-mixing' the dimensionless Dispersion Number has values of 0 and ∞ respectively. All real systems have values between these two extremes.

Chudoba's results, which were obtained on pilot plants treating synthetic sewage, have been confirmed on full-scale activated-sludge plants by Tomlinson and Chambers [7]. Here the Dispersion Number was correlated with Stirred Specific Volume Index (SSVI) as recommended by White [8]. Plants exhibiting low Dispersion Numbers, that is those tending towards ideal 'plug-flow', produced sludges with better settling characteristics than those which exhibited high Dispersion Numbers and approached 'complete-mixing' conditions. The data describing the relationship showed considerable scatter. Apart from experimental errors which occurred during the measurement of Dispersion Numbers at the plants in question, it was recognized that other factors relating to plant design and waste water characteristics could affect sludge settleability. The WRC has made further investigations on the effects of mixing characteristics on settleability of sludge in pilot-scale plants. This chapter describes the results of those experiments.

10.2 EXPERIMENTAL PROCEDURES AND RESULTS

10.2.1 The use of anoxic mixing zones to control settleability

Experiments were performed by incorporating anoxic zones into two different types of activated-sludge plant. The first had an aeration tank consisting of a single compartment and could be considered as a 'complete-mixing' system. The second plant had an aeration tank consisting of eight equal-sized compartments connected in series. This tank represented an approach to 'plug-flow' characteristics.

The average operating conditions of both plants are shown in Table 10.1.

Table 10.1 – Operating conditions of pilot-scale activated-sludge plants

	Single compartment 'complete-mixing' plant	8 compartment 'plug-flow' plant
Aeration tank volume (m^3)	0.38	0.1
Settled sewage flow rate (l/h)	16	12
Sludge loading (kg BOD/kg MLSS day)	0.12	0.2
Sludge age (d)	20	18

10.2.1.1 Single-compartment 'complete-mixing' plant

Anoxic conditions were created in this plant by installing a removable mixing tank at the inlet to the aeration tank. Settled sewage and recycled sludge flowed into this tank and solids were kept in suspension by the action of a low-speed mechanical stirrer. The results obtained during an 8-month period of operation are shown in Fig. 10.1.

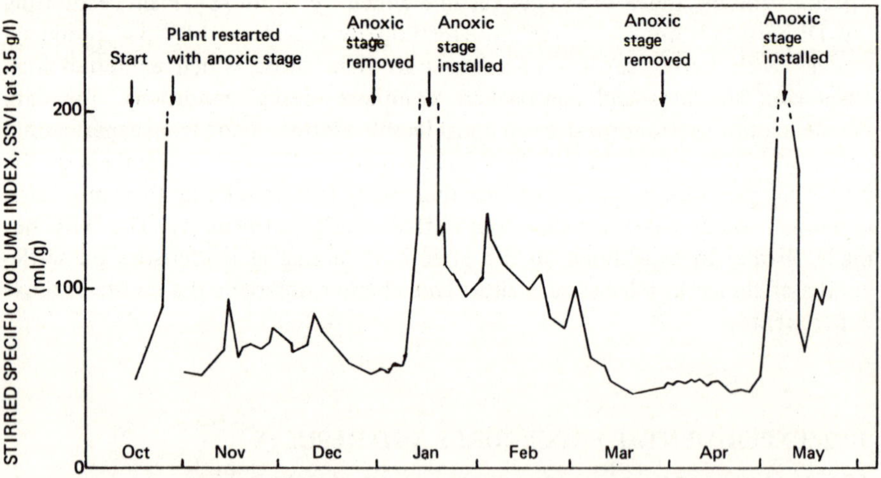

Fig. 10.1 – Effect of anoxic stage on the settleability of activated sludge in a 'completely-mixed' plant.

It was known at this time that the treatment of settled domestic sewage from Stevenage in a single-compartment tank often resulted in sludges with very poor settling characteristics. Thus the plant was originally started, without the anoxic stage, with a nitrifying, non-filamentous activated sludge with good settling properties, brought in from an external source. Bulking soon occurred and sludge was lost from the system. The plant was restarted again with the non-filamentous sludge, but this time the anoxic mixing stage was incorporated. For a period of two months the plant operated in a stable manner with the Stirred Specific Volume Index (at 3.5 g/l) remaining below 100 ml/g. At the end of this period, the anoxic stage was removed and the settleability deteriorated rapidly. Before serious loss of solids occurred, the anoxic stage was reinstalled and then, over a period of two months, the settleability improved until it reached its original value. Finally the anoxic stage was again removed, and after a deterioration in settleability it was reinstalled. Again this brought about an improvement in settleability.

10.2.1.2 Eight-compartment 'plug-flow' plant

Two plants were used in these experiments. Each consisted of eight compartments in series but in one plant the contents of the first compartment were stirred mechanically to form an anoxic denitrification stage, whilst in the other plant all compartments were aerated.

The results obtained during a six-month period of operation are shown in Fig. 10.2. The 'conventional' plant refers to the plant without the anoxic stage. Approximately every six weeks the settleability of the sludge in the conventionally-operated plant deteriorated to such an extent that it had to be replaced. However, settleability of the sludge in the modified plant remained at a satisfactory value throughout most of the period. During October a further anoxic stage was incorporated into the modified plant by operating the second compartment anoxically also. This resulted in the remainder of the aeration tank being overloaded and a partial loss of nitrification occurred, which was associated with the deterioration of settleability during this period.

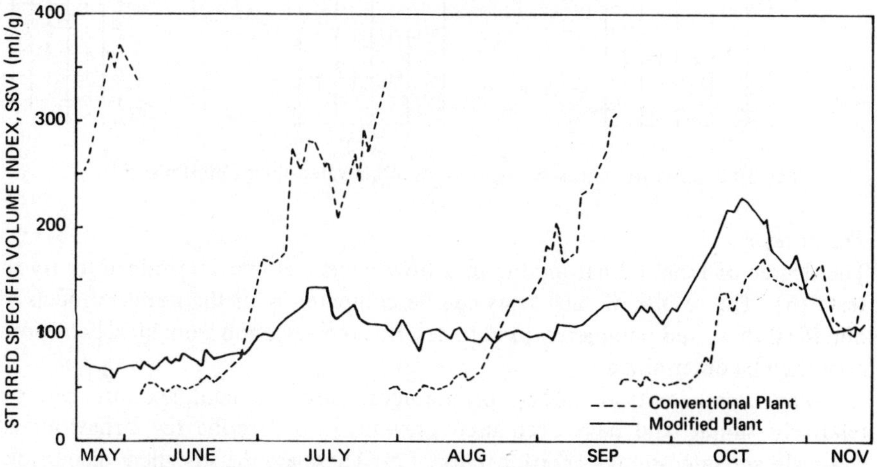

Fig. 10.2 – Settleability of sludge from 'plug-flow' plants.

10.2.2 The effect of longitudinal mixing on sludge settleability

Experiments were performed in seven activated-sludge pilot plants, each having an aeration tank volume of 1 m^3. Six of the plants were designed for continuous operation in the conventional manner, the seventh was constructed as a batch fed 'fill-and-draw' system.

10.2.2.1 Conventional plants

These plants were identical to those previously described by Tomlinson and Chambers [4], except for the design of the aeration tanks. A range of flow conditions between 'complete-mixing' and 'plug-flow' was achieved by dividing the aeration tanks into numbers of equal-sized compartments in series. For a small number of compartments it is convenient to insert baffles into the aeration tank. Greater numbers of compartments were found to be more easily constructed from separate modules connected by short lengths of piping. Aeration tanks having 1, 2, 4, 8, 12 and 24 compartments in series were constructed. A diagrammatic representation of the tank configuration is shown in Fig. 10.3.

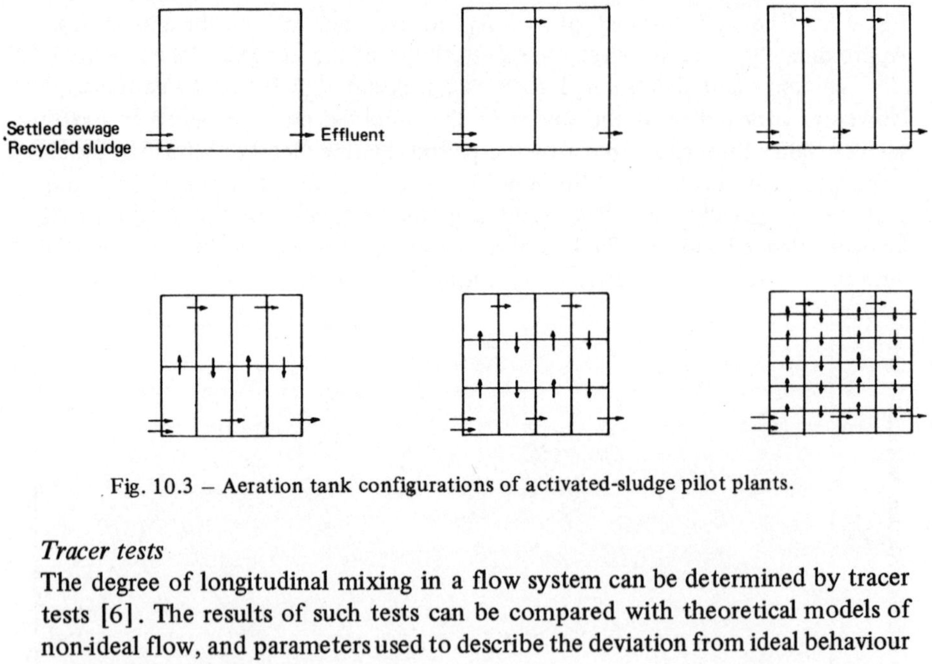

Fig. 10.3 – Aeration tank configurations of activated-sludge pilot plants.

Tracer tests

The degree of longitudinal mixing in a flow system can be determined by tracer tests [6]. The results of such tests can be compared with theoretical models of non-ideal flow, and parameters used to describe the deviation from ideal behaviour may then be determined.

Many mathematical models of non-ideal flow are complex but two are relatively simple and have been used previously to describe the behaviour of full-scale activated-sludge aeration tanks [7]. These are the so-called 'dispersion' and 'tanks in series' models.

The 'dispersion' model assumes longitudinal mixing occurs in a manner analogous to diffusion processes. It is possible to quantify non-ideal flow using this model in terms of a dimensionless Dispersion Number, which varies from nought for 'plug-flow' to infinity for 'complete-mixing'.

The 'tanks in series' model assumes that non-ideal flow may be represented by a system consisting of an appropriate number of equal-sized 'complete-mixing' tanks. Here the degree of longitudinal mixing is dependent on the number of tanks in the series. The number of tanks can vary between one for 'complete-mixing' and infinity for 'plug-flow'.

It was decided to employ the 'tanks in series' model to describe the longitudinal mixing in the pilot plant aeration tanks as it is less involved mathematically than the dispersion model. Moreover, the aeration tanks were in reality divided into numbers of 'completely-mixed' tanks in series and so it was relatively simple to test that the model provided an accurate description of the flow behaviour.

Tracer tests were performed on the aeration tanks previously described and the experimental results were compared with the theoretical predictions of the 'tanks in series' model. Tracer material was injected into the inlet of the aeration tanks in the form of a 'pulse'. The response of a flow system, measured at the outlet, to such an input, takes the form of a tracer concentration versus time curve which is a residence time distribution of fluid in the tank.

The response of a series of ideal 'completely-mixed' tanks to a pulse input of tracer material is a distribution of tracer concentration versus time described by the following equation:

$$C = \frac{C_t}{C_o} = \frac{N^N}{(N-1)!} \theta^{N-1} e^{-N\theta} \tag{10.1}$$

The relative mathematical simplicity of the 'tanks in series' model facilitates direct comparison of experimental tracer output results with the theoretical curves of Equation 10.1. The results of this procedure are shown in Fig. 10.4.

A number of other methods may be used to obtain a value of N from an experimental tracer curve. If Equation 10.1 is differentiated with respect to θ then it can be shown that the maximum value of each distribution occurs at

$$\theta = \frac{N-1}{N} \tag{10.2}$$

An estimate of N may therefore be made by inspection of the maximum on an experimental tracer output curve.

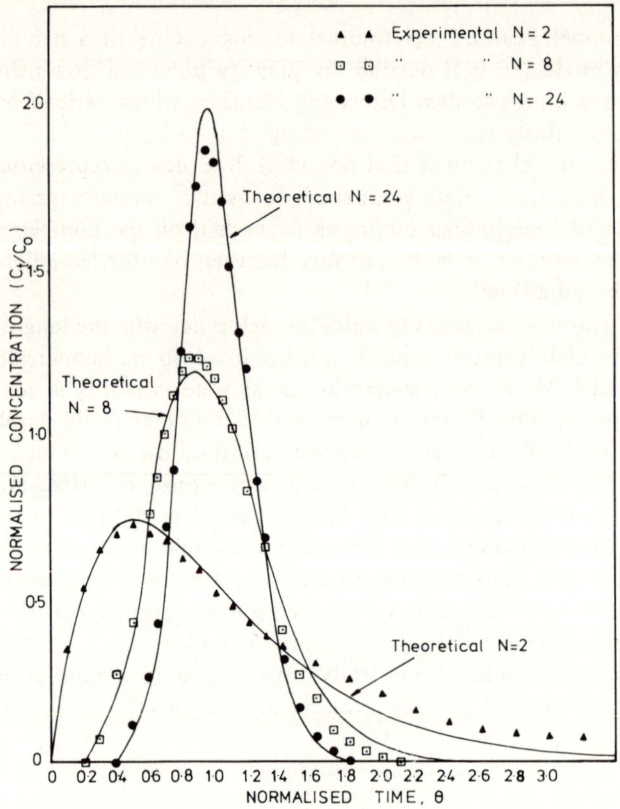

Fig. 10.4 – Theoretical and experimental residence time distributions.

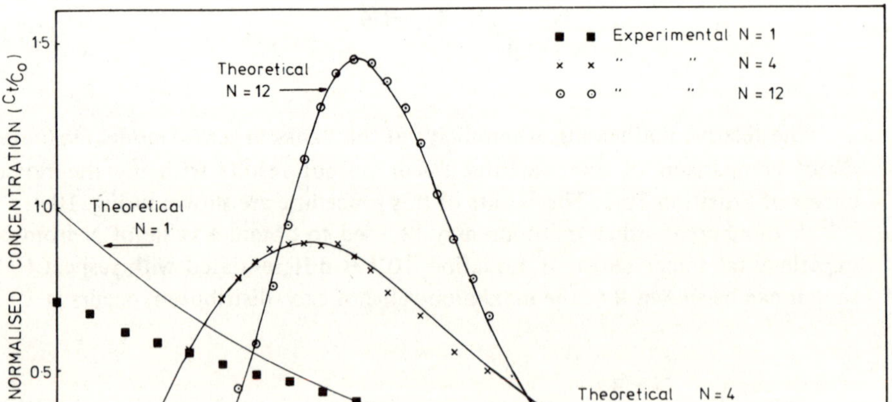

The value of N may also be estimated by calculating the variance, σ^2, of the experimental distribution. It is possible to show that the variance of Equation 10.1 is given by:

$$\sigma^2 = \frac{1}{N} \qquad (10.3)$$

Substituting the variance of the experimental curve into Equation 10.3 allows N to be determined.

Values of N calculated from the experimental tracer curves as described above are compared with the actual number of tanks in series in each aeration tank in Table. 10.2. This table also shows, for purposes of illustration, the Dispersion Numbers of each aeration tank.

Table 10.2 – Comparison of experimental values of N with actual number of compartments in each aeration tank

N actual	N from Eqn. 10.2	N from Eqn. 10.3	Dispersion Number
1	1	1.5	0.720
2	2	2.6	0.264
4	4	4.4	0.128
8	8.5	7.7	0.053
12	12.8	12.2	0.041
24	25	23	0.023

It was concluded from Fig. 10.4 and Table 10.2 that the 'tanks in series' model described fairly accurately the flow conditions in the pilot-plant aeration tanks. The single-compartment aeration tank could be considered as an approximation to a 'complete-mixing' system while the 24-compartment tank was a reasonable practical approach to 'plug-flow' conditions.

The effect of sewage septicity on the settleability of sludge

Poor sludge settleability has been associated with sewage septicity since the earliest days of the use of the activated-sludge process [9, 10]. Recently a survey of activated-sludge plants in the UK revealed that several received septic sewage, but that treatment of septic sewage did not necessarily cause bulking in all cases [1]. It was decided to perform pilot-scale experiments to investigate the effect of sewage septicity on sludge settleablity, using the pilot plants previously described.

The distinction between 'fresh' and 'septic' sewage is somewhat arbitrary. Sewage which reaches a treatment works after passage through rising main sewers or enters an aeration tank after long periods of retention in primary or storm tanks, is often anaerobic. Analysis of such sewage reveals the presence of sulphide and appreciable concentrations of intermediate anaerobic reaction products such as acetic, propionic and other volatile organic acids.

At the Stevenage Laboratory of the WRC it was possible to obtain a supply of both fresh and septic sewage. The fresh sewage was intercepted from a gravity sewer carrying sewage from a purely domestic source. Septic sewage was obtained by allowing fresh sewage to remain in storage tanks for several days before use. Analyses performed on fresh and septic sewage samples gave the average results shown in Table 10.3.

The six activated-sludge pilot plants were used to treat fresh sewage for an operating period of about three months. The plants were then restarted using septic sewage and operated for a further period of about three months. Apart from the different sewage feeds, other operating parameters were maintained at approximately the same values during the two experimental periods. In each case sewage and recycled sludge flow rates were set at 100 l/h, giving a mean hydraulic residence time in each aeration tank of 5 hours.

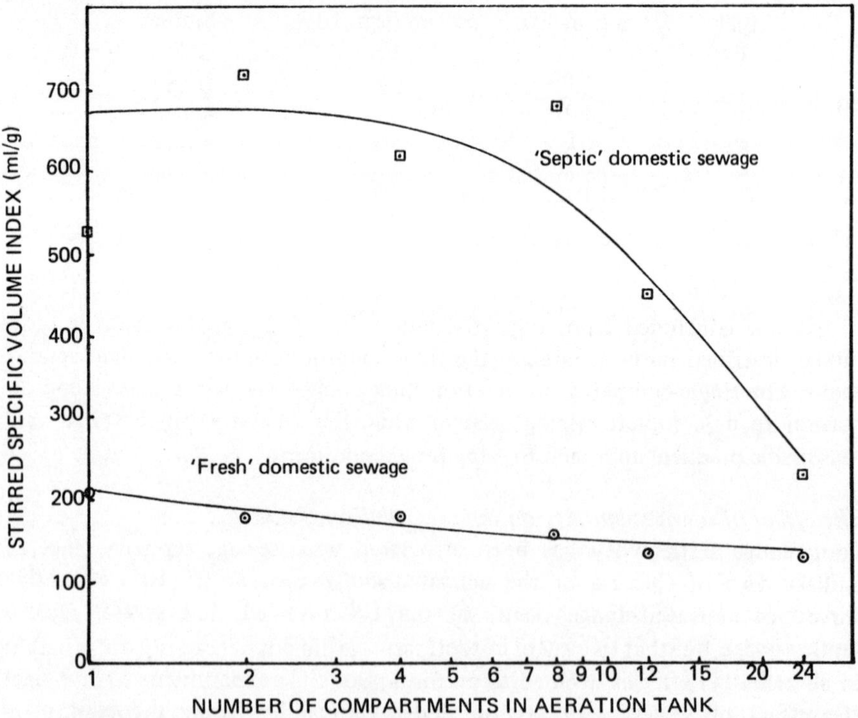

Fig. 10.5 – Settleability of sludge from pilot plants treating 'fresh' and 'septic' domestic sewage.

The results obtained are shown in Fig. 10.5 as a graph of sludge settleability versus the number of compartments in the aeration tank for the treatment of fresh and septic domestic sewage. It can be seen that the settling characteristics of the sludges produced in the plants treating septic sewage were always far worse than in the corresponding plants treating fresh sewage. Only the 24-compartment aeration tank could produce a sludge with an average SSVI below 250 ml/g when treating septic sewage. Bulking was so severe in the other plants treating septic sewage that the MLSS could not be maintained at a concentration above about 1500 mg/l and carbonaceous oxidation was reduced.

Table 10.3 – Analysis of fresh and septic sewage feed to pilot-scale activated-sludge plants

	Fresh sewage	Septic sewage
Sulphide (mg/l)	<0.5	15
†Volatile organic acids (mg/l)	45–50	100–150

†Mainly acetic with smaller concentrations of propionic, butyric and valeric acids.

The effect of recycle ratio and hydraulic residence time on the settleability of activated sludge

Although the effect of longitudinal mixing on sludge settleability is well established, the reason why 'plug-flow' plants tend to produce sludges with better settling characteristics than those from 'completely-mixed' plants remains unclear. In a 'completely-mixed' system all the sewage which enters the aeration tank is immediately dispersed and mixed with the rest of the tank contents. The microorganisms experience a uniform low level of substrate concentration. In a 'plug-flow' plant incoming sewage is not dispersed, as there is no longitudinal mixing and consequently the micro-organisms are initially surrounded by a much higher concentration of substrate. This substrate is gradually utilized as the mixed liquor flows through the aeration tank towards the outlet. It is thought that the existence of such a substrate concentration gradient is important in controlling the settling characteristics of the sludge.

In a conventional plant, changes in the recycled sludge flowrate can modify this gradient and high recycle rates can effectively change the characteristics of a 'plug-flow' plant into those of a 'completely-mixed' system. The recycling of sludge at high rates introduces large quantities of treated effluent into the aeration tank inlet and the aeration tank environment becomes more uniform. Pilot plant experiments have been performed to investigate the effect of recycle ratio on sludge settleability.

In a full-scale activated-sludge plant changes in the recycle ratio will normally produce corresponding increases or decreases in the hydraulic retention period since this variable is a function of recycle and sewage flowrates, and the sewage flowrate is not usually under direct control. In order to differentiate between the effect of recycle ratio and hydraulic residence time on sludge settleability, a series of pilot-scale experiments were conducted at constant hydraulic residence time and different recycle ratios, and at constant recycle ratio with different hydraulic residence times. The operating conditions employed and the settleability of the sludges produced during the experiment are given in Table 10.4.

Table 10.4 – Operating conditions and sludge settleabilities in pilot plants at various recycle ratios and hydraulic residence times

				SSVI (at 3.5 g/l) Number of compartments					
q_s (l/h)	q_r (l/h)	r	$\bar{t}$ (h)	1	2	4	8	12	24
114	86	0.75	5.0	213	160	170	142	143	105
100	100	1.0	5.0	210	180	182	161	140	131
89	111	1.25	5.0	205	183	184	162	137	135
150	150	1.0	3.3	273	212	175	163	148	102
63	63	1.0	8.0	90	80	99	89	85	87

The settling characteristics of the sludges produced in the pilot plants are given as SSVI (at 3.5 g/l) in Table 10.4. These values are averages calculated from many determinations made during about three months of operation under the conditions shown.

The results obtained are plotted as graphs of settleability against the number of compartments in the aeration tank in Figs 10.6, 10.7 and 10.8. It can be seen that the plants with aeration tanks divided into many compartments, that is those tending towards 'plug-flow' produced sludges with better settling characteristics than those systems which tended towards 'complete-mixing'. This effect was evident under all the conditions of operation employed, but was much more marked in some cases than in others. Thus at long hydraulic residence times the effect of longitudinal mixing on settleability is slight and all the plants produced sludges with good settling characteristics as shown in Fig. 10.8. At short hydraulic residence times, however, the plants which tended towards 'complete-mixing' produced sludges with poor settling characteristics in comparison to the plants which exhibited higher degrees of 'plug-flow'. The general form of the relationship between settleability and longitudinal mixing is shown in Fig. 10.6. Figure 10.7 shows the effect of varying the recycle ratio, at constant hydraulic time, on sludge settleability. It can be seen that changes in the recycle ratio, within the limits employed, do not have a great effect on sludge settleability providing the hydraulic residence time remains constant.

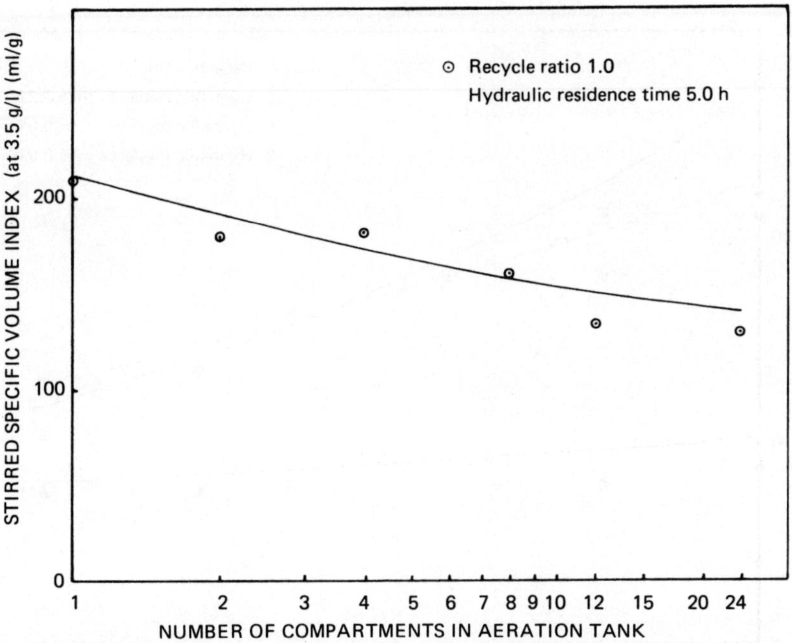

Fig. 10.6 – Settleability of activated sludge from pilot plants exhibiting various degrees of longitudinal mixing.

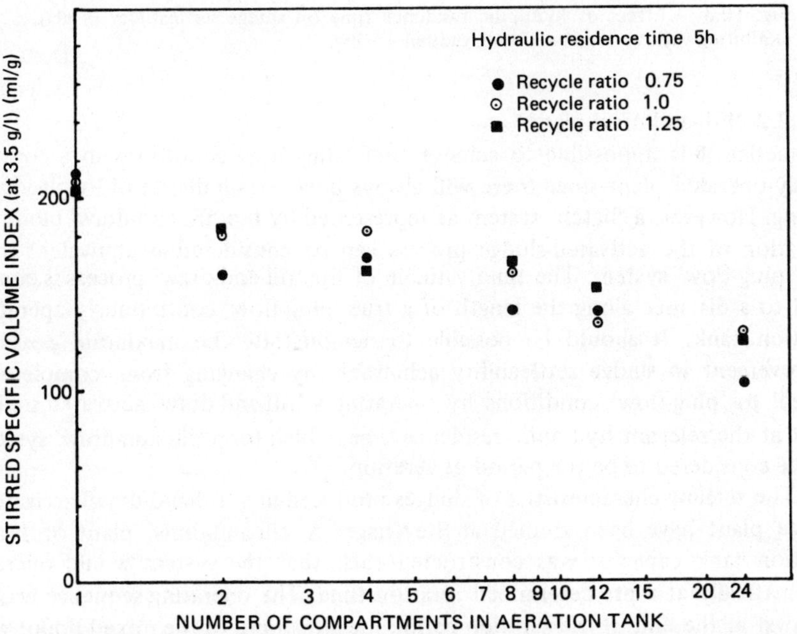

Fig. 10.7 – Effect of recycle ratio on sludge settleability in plants exhibiting various degrees of longitudinal mixing.

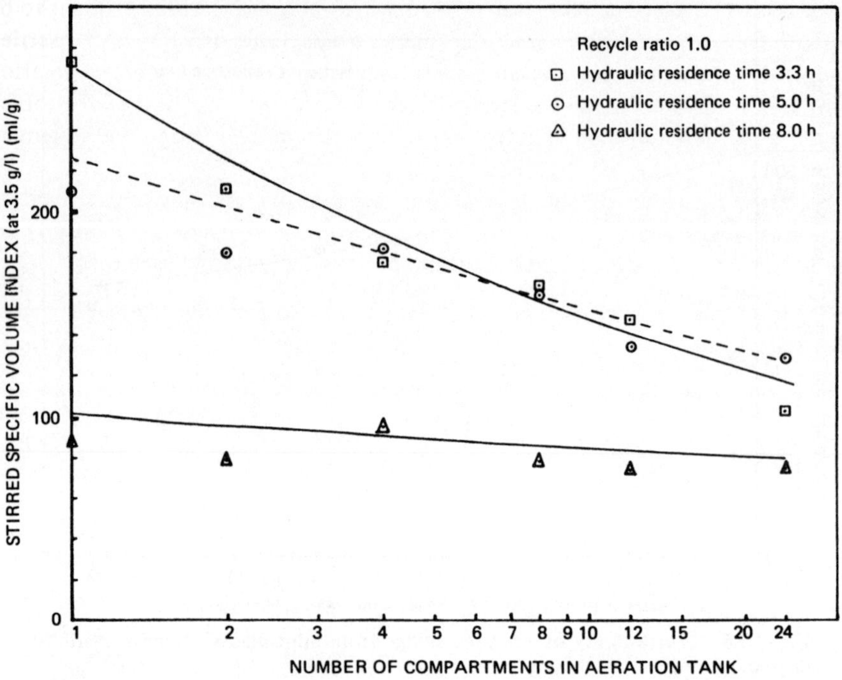

Fig. 10.8 – Effect of hydraulic residence time on sludge settleability in plants exhibiting various degrees of longitudinal mixing.

10.2.2.2 'Fill-and-draw plant

In practice it is impossible to achieve true 'plug-flow' conditions in a continuously-operated plant since there will always be a certain degree of longitudinal mixing. However, a 'batch' system as represented by the 'fill-and-draw' mode of operation of the activated-sludge process can be considered as equivalent to a true 'plug-flow' system. The time variable of the 'fill-and-draw' process is equivalent to a distance along the length of a true 'plug-flow' continuously-operated aeration tank. It should be possible to demonstrate the maximum possible improvement in sludge settleability achievable by changing from 'completely-mixed' to 'plug-flow' conditions by operating a 'fill-and-draw' activated-sludge plant at the relevant hydraulic residence time, which for a 'fill-and-draw' system can be considered to be the period of aeration.

The settling characteristics of sludges produced in a 'fill-and-draw' activated-sludge plant have been studied at Stevenage. A 'fill-and-draw' plant of 1 m^3 aeration tank capacity was constructed such that the system would operate automatically at a pre-determined aeration time. The operating sequence was as follows: at the end of the aeration period the air supply to the mixed liquor was cut off and the suspended solids were allowed to settle for half an hour. After the settlement period a valve was opened which allowed clarified effluent to be run off from the aeration tank. The inlet to this valve was above the settled sludge/effluent interface. The air supply was then turned on and the aeration tank was refilled to a pre-set level with settled sewage. Aeration of the mixed liquor was then resumed. Sludge was removed manually from the system as necessary to keep the MLSS concentration at about 2.5 to 3 g/l.

The settleability of the sludges produced in the 'fill-and-draw' plant at various aeration times are shown in Fig. 10.9. The SSVI values (at 3.5 g/l) are averages obtained during a three-month experimental period in each case.

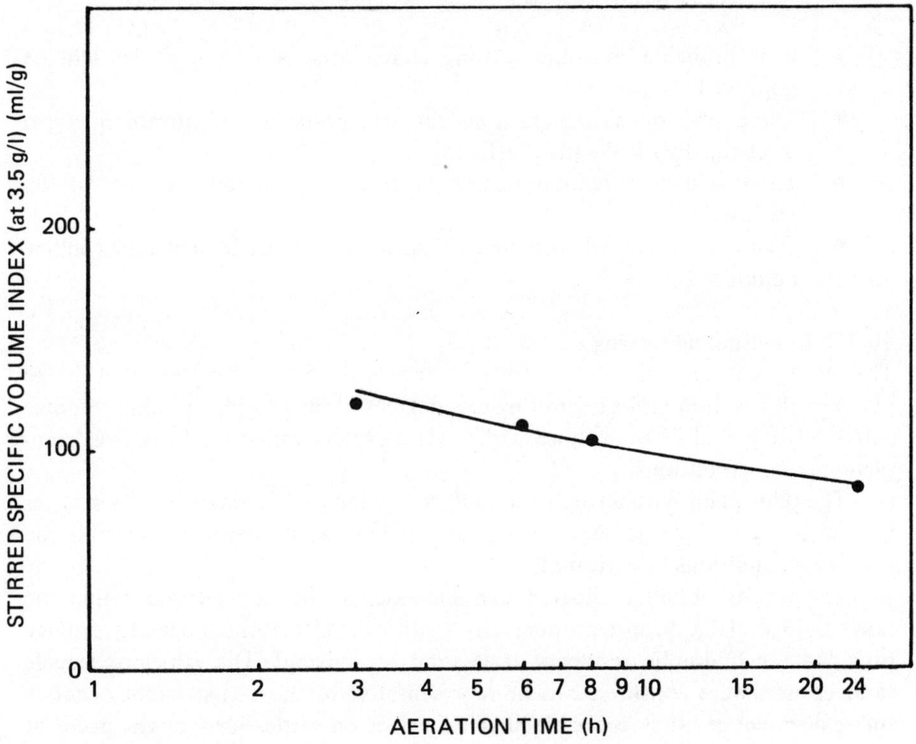

Fig. 10.9 – Effect of aeration time on sludge settleability from 'fill-and-draw' plant.

10.3 DISCUSSION OF RESULTS

10.3.1 The anoxic mixing stage

At this time it is uncertain why an anoxic environment brings about an improvement in sludge settling characteristics. Some effect is probably due to the decrease in longitudinal mixing. The hydraulic retention time in the anoxic zones used in the reported experiments was about 25 to 30 minutes. During this period it was found that a substantial proportion of the nitrate in the recycled sludge was utilized as an oxygen source by denitrifying organisms. It is possible that the selection of microbial flora is affected by the denitrification reactions which occur, or by the adsorption of substrate by organisms in the absence of normal aerobic metabolism.

In the anoxic zone oxygen used in the aeration tank to convert ammonia to nitrate is recovered and utilized. This oxygen represents a small but significant proportion of the overall oxygen requirements of the process. The incorporation of an anoxic stage into a nitrifying plant has the following advantages:

- It is probable that the settling characteristics of the sludge will be improved.
- There will be a reduction in the concentration of nitrate nitrogen discharged with the plant effluent.
- There will be a reduction in the overall oxygen requirements of the process.
- Problems associated with denitrification in final settlement tanks will be minimized.

10.3.2 Longitudinal mixing

The pilot-plant experiments performed have confirmed that activated-sludge plants with aeration tanks exhibiting low degrees of longitudinal mixing produce better settling sludges than those with aeration tanks which tend towards 'complete-mixing' conditions.

The pilot plant with an aeration tank consisting of 24 compartments in series was able to produce sludges with good settling characteristics under all the operating conditions investigated.

The results obtained showed that increases in the recycle ratio within the range 0.75 to 1.25 do not produce any significant effect on sludge settleability provided the hydraulic residence time remains constant. The values of recycle ratio chosen were considered to be representative of those that might occur in full-scale practice. It is probable that an effect on settleability might occur at much higher recycle ratios. Experiments performed at recycle ratios of less than 0.75 produced sludges which did not flocculate and a corresponding poor plant performance. The results of such trials are not presented. However, it is known that some full-scale plants operate satisfactorily at recycle ratios below 0.75 and it is also possible that some effect on settleability may occur at these low values.

At the inlet to an activated-sludge aeration tank the mixing of the influent sewage and the recycled sludge results in an instantaneous substrate concentration which is lower than that in the sewage feed. This concentration, C_i, is given by the expression

$$C_i = \frac{C_f + r\,C_e}{1 + r} \qquad (10.4)$$

where C_f is the concentration of substrate in the sewage,
C_e is the concentration of substrate in the effluent, and
r is the recycle ratio.

If C_f remains constant and it is assumed that C_e also remains constant, it can be seen that as r increases then C_i decreases and eventually an approach will be made to the situation where $C_i \cong C_e$. When the latter situation prevails the concentration gradient in the aeration tank will no longer exist and the uniform environment of complete mixing will result. However, it is easily shown that normal values of the recycle ratio do not result in a substantial modification of the concentration gradient. If, in Equation 10.4, C_f is taken to be 250 mg/l (measuring the substrate as BOD) and C_e is 20 mg/l, then for r = 1.0, C_i is 135 mg/l. If r is increased to 2.0, C_i is only reduced to about 97 mg/l.

It is unlikely, in full-scale practice, to employ recycle ratios in excess of about 2.0 and it might be expected that recycle ratios of less than the latter value should not affect the settling characteristics of the sludge at constant hydraulic residence time. The results shown in Fig. 10.7 confirm this hypothesis for recycle ratios in the range 0.75 to 1.25.

Normally, in a full-scale plant, increases in the recycle ratio result in decreases in hydraulic residence time, since the sewage influent flow will not be under direct control. This is evident from the equation

$$\bar{t} = \frac{V}{q_s\,(1 + r)} \qquad (10.5)$$

where q_s is the sewage influent flowrate, and $\bar{t}$ is the hydraulic residence time.

The effect of longitudinal mixing on sludge settleability at various aeration-tank hydraulic residence times, as shown in Fig. 10.8, is difficult to explain in terms of substrate concentration gradients. At a residence time of 8.0 h the establishment of a concentration gradient in the aeration tank did not appear to be important in controlling the sludge settleability. However, at shorter residence times, only those plants with aeration tanks exhibiting high degrees of 'plug-flow' could produce sludges with good settling characteristics. It would appear that for a 'completely-mixed' plant, the settleability deteriorates with increasing loading. This effect is the opposite to that established by Chudoba *et al.* [11] and suggests a maximum value somewhere on the curve of settleability versus loading. It is difficult to quantify the effect of sludge loading on settleability in a conventional plant because when bulking is severe the MLSS concentration is reduced and the loading may change from the desired value.

The results obtained in the 'fill-and-draw' plant show that the settleability of the sludges produced was in general comparable with the settleability of the sludges produced in the 24-compartment continuously-operated plant. It is probable, therefore, that no substantial improvement in sludge settleability will result from the operation of a plant with a higher degree of 'plug-flow' than is represented by a series of 24 well-mixed compartments.

10.4 CONCLUSIONS

Pilot plant investigations conducted at the Stevenage Laboratory of the WRC have shown that:

- The incorporation of an anoxic zone into a conventional activated-sludge plant achieving full nitrification can have a beneficial effect on the settleability of the sludge in that plant.
- The treatment of 'septic' sewage in an activated-sludge plant results in sludges with poor settling characteristics in comparison with 'fresh' sewage.
- Plants which tend towards ideal 'plug-flow' produce better settling sludges than those which exhibit a high degree of longitudinal mixing.
- Plants which tend towards ideal 'plug-flow' produce sludges with good settling characteristics at aeration tank hydraulic residence times in the range 3.3 to 8.0 h. Plants exhibiting a high degree of longitudinal mixing can only produce sludges of good settleability at high values of aeration tank residence time.
- The effect of recycle ratio on sludge settleability is slight providing the recycle ratio is in the range 0.75 to 1.25 and providing the hydraulic residence time remains constant.
- A 'fill-and-draw' activated-sludge plant produces sludge with settling characteristics similar to those from continuously-operated plants exhibiting a high degree of 'plug-flow'.

10.5 NOMENCLATURE

C	Dimensionless tracer concentration C_t/C_o
C_t	Concentration of tracer at time t (kg/m^3)
C_o	Initial tracer concentration (kg/m^3)
θ	Dimensionless time $t/\bar{t}$
e	Exponential function
N	Number of ideal equal-sized completely-mixed tanks in series to which a flow system is equivalent
σ^2	Variance of distribution function
q_s	Settled sewage flowrate (l/h)
q_r	Recycled sludge flowrate (l/h)
r	Recycle ratio $\frac{q_r}{q_s}$

$\bar{t}$	Hydraulic residence time (h)
C_f	Concentration of substrate in settled sewage (mg/l)
C_e	Concentration of substrate in effluent (mg/l)
C_i	Instantaneous concentration of substrate at aeration tank inlet (mg/l)
V	Aeration tank volume (l)

10.6 ACKNOWLEDGEMENTS

The author would like to express his appreciation of the work performed during these investigations by Messrs T. Bennett and J. P. Ryan, Sandwich-course students from the Chemical Engineering Department of Loughborough University.

10.7 REFERENCES

[1] Tomlinson, E. J. *Bulking – A Survey of Activated Sludge Plants.* Technical Report TR 35, Water Research Centre, 1976, 32 pp.

[2] Cooper, P. F., Collinson, B. and Green, M. K. Recent advances in sewage effluent denitrification. Part II. *Water Pollution Control,* 1977, **76**, 389–401.

[3] Heide, B. A. and Pasveer, A. Oxidation ditch: prevention and control of filamentous sludge. *H_2O*, 1974, **7**, 373–377.

[4] Tomlinson, E. J. and Chambers, B. *The use of Anoxic Mixing Zones to Control the Settleability of Activated Sludge.* Technical Report TR 116, Water Research Centre, 1979, 24 pp.

[5] Chudoba, J., Ottová, V. and Maděra, V. Control of activated sludge filamentous bulking – I. Effect of the hydraulic regime or degree of mixing in an aeration tank. *Water Research,* 1973, **7**, 1163–1182.

[6] Levenspiel, O. *Chemical Reaction Engineering,* 1962, John Wiley and Sons, New York and London.

[7] Tomlinson, E. J. and Chambers, B. *The Effect of Longitudinal Mixing on the Settleability of Activated Sludge.* Technical Report TR 122, Water Research Centre, 1979, 46 pp.

[8] White, M. J. D. *Settling of Activated-Sludge.* Technical Report TR 11, Water Research Centre, 1975, 27 pp.

[9] Smit, J. Bulking of activated sludge – II, Causative organisms. *Sewage Works Journal,* 1934, **6**, 1041.

[10] Greeley, S. A. The development of the activated sludge method of sewage treatment. *Sewage Works Journal.* 1945, **17**, 1135.

[11] Chudoba, J., Bláha, J. and Maděra, V. Control of activated sludge filamentous bulking – III Effect of sludge loading. *Water Research,* 1974, **8**, 2311–1237.

Discussion

Chairman: **G. S. Clapham** (Wessex Water Authority, UK)

Professor P. Grau (Prague Institute of Chemical Technology, Czechoslovakia) opened the discussion and, with the aid of diagrams, reiterated a point made by the author that any real tank exhibits some degree of dispersion or backmixing and some degree of plug-flow. Further, he described how a real tank could be simulated hydraulically by a series of ideal tanks each of which was itself fully dispersed or backmixed but did not backmix with other tanks in the series. He said that this approach of characterizing real tanks in terms of an equivalent number of ideal tanks in series was one which he supported fully, since other methods of characterization, such as measurement of Dispersion Numbers (D/uL), were sometimes difficult to visualize in real terms. Professor Grau then presented the following equation which he said could be used to estimate the 'equivalent number of ideal tanks in series', j, from the Dispersion Number, D/uL

$$j = 0.5/\frac{D}{uL} + 0.7$$

Mr. E. J. Tomlinson (Water Research Centre) continued the discussion and explained why the number, j, was not necessarily equal to the actual number of tanks. In laboratory-scale equipment mixing is generally very efficient and j is equal to the actual number of tanks. However, in full-scale activated-sludge plants mixing is not as efficient and one actual tank may be equivalent to more than one ideal tank. He explained that his own analysis had shown that to achieve good settleability of sludge it is necessary to design aeration tanks with j values of 10 to 15 or more. Unfortunately, it is only possible to determine the j value of a tank by tracer studies and these cannot be performed until the tank has been constructed. Therefore, he invited delegates to suggest methods for prediction of values of j at the design stage. **Professor Grau** replied that he had a method based upon tank size and velocity of flow between tanks.

In relation to the measurement of mixing characteristics **Mr. G. McBrearty** (Thames Water Authority, UK) asked the author what type of tracer had been used for the investigations and at what point its concentration was measured. Further, he wondered what forms of cheap baffling material could be employed to increase the degree of plug flow exhibited by aeration tanks. **Dr. Chambers** replied that the tracer employed most commonly was lithium chloride, though any element which could be easily detected and was not already present in large or irregular concentrations in sewage could be employed. In some cases, where the quantity of lithium chloride would have been excessively large a radiotracer (Br–82) had been used. He pointed out that radiotracers, although requiring handling by specialists, were extremely convenient because continuous instrumental monitoring could be employed. He confirmed that measurement of tracer concentration is made at the outlet of the aeration tank as it is the characteristics of this tank which are being measured. **Professor Grau** told delegates that in his laboratory also, lithium chloride was used but in some cases he found it useful to employ a dyestuff so that the test could be followed visually, or sodium chloride so that measurements of conductivity could be made. With regard to the construction of baffles, **Dr. Chambers** said that various materials had been employed. He did not think that canvas tarpaulin was adequate but felt that marine ply-wood would be adequate in many cases.

Professor H. Verachtert (University of Louvain, Belgium) asked the author how he would explain the apparent contradiction that septic sewage caused deterioration of settleability whilst the incorporation of an anoxic zone improved it. **Dr. Chambers** stressed that septic conditions did not exist in the anoxic zone because oxygen was available in the form of nitrate. Therefore, he did not think that the evidence was contradictory. Continuing on the question of anoxic zones, **Dr. J. Houtmeyers** (Forest Biology Centre, Belgium) asked whether the benefits obtained might not be a result of the reduction of longitudinal mixing. **Dr. Chambers** agreed that this probably accounted for some of the improvement, though a series of experiments in which nitrifying and non-nitrifying plants were operated with aerated or unaerated premixing tanks had indicated that for nitrifying plants some considerable benefit was obtained from the anoxic conditions. In reply to Mr. McBrearty **Dr. Chambers** said that neither anoxic zones nor plug-flow conditions would overcome problems arising as a result of nitrogen or phosphorus deficiency.

There was then some discussion about the ability of anoxic zones to reduce the proportion of filamentous organisms in activated sludge. **Dr. Chambers** told Professor Verachtert that work at the Water Research Centre had failed to identify any clear and definite relationship between settleability and presence of specific filamentous organisms and that improvements in settleability by incorporation of anoxic zones had been achieved without noticeable change in the content of filamentous organisms. Therefore, he saw no reason for regular microbiological monitoring of sludge for operational purposes and preferred the quicker and simpler settleability test. **Dr. C. Forster** (University of Birmingham, UK) asked whether filamentous organisms were therefore irrelevant and whether the surface characteristics of the sludge were more important. **Dr. Chambers** disagreed. Replying to Dr. Chambers comments on the value of regular microbiological monitoring **Mr. D. H. Eikelboom** (Research Institute for Environment Hygiene, TNO, The Netherlands) suggested that such examinations were much less time consuming that settlement tests.

Chapter 11

The use of chlorination for control of activated sludge bulking

D. Jenkins, J. B. Neethling, H. Bode, and **M. G. Richard,** University of California, Berkeley, USA

11.1 INTRODUCTION

In the activated-sludge treatment of domestic waste water and many industrial wastes, secondary sedimentation is often the unit process that determines BOD and suspended solids (SS) removal efficiency. Types of solids separation problems encountered are summarized in Table 11.1. This chapter deals with our approach to the practical control and resolution of bulking problems.

Table 11.1 – Activated Sludge Solids Separation Problems

Name of Problem	Nature of Problem	Manifestation of Problem
Dispersed growth	Micro-organisms do not form flocs but are dispersed, forming only small clumps or single cells	Turbid effluent. No floc formation. No zone settling of sludge
Pin floc or Pinpoint floc	Small, compact, roughly spherical flocs are formed, the larger of which settle rapidly leaving smaller ones behind to form a turbid supernatant	Low SVI –cloudy, turbid effluents
Bulking	Filamentous organisms extend from flocs into bulk solution and interfere with compaction and settling of activated sludge	High SVI – very clear supernatant. Low RAS solids concentration. In severe cases, loss of process control due to sewage blanket overflowing secondary clarifier weir and loss of efficiency of solids handling processes
Blanket rising	Denitrification in secondary clarifier releases N_2 bubbles which attach to activated-sludge patches floating them to secondary clarifier surface	Gas bubbles rise to surface of secondary clarifier. Scum of activated sludge forms on surface of secondary clarifier
Foaming	Foams caused both by non-degradable surfactant and by *Nocardia* sp.	Foams float large amounts of activated sludge solids to surface of treatment units. *Nocardia* foams are persistent and difficult to break mechanically

11.2 BULKING CONTROL APPROACH

A two-stage approach is recommended. The first stage is a non-specific treatment that eliminates bulking symptoms by selectively killing off filamentous micro-organisms extended from the flocs. In the second stage, the causes of filamentous organism growth are determined and remedied. A two-stage approach is needed because:

- Bulking has many causes, some of which are poorly understood. Thus immediate specific control measures may not be possible.
- Rapid bulking control is often required. Following a change in conditions to disfavour filamentous organisms growth, approximately three mean-cell residence times (MCRT) are required for activated sludge to return to a non-filamentous condition [1]. For a plant with an MCRT of ten days, one month would be required.

The first stage is to apply controlled doses of a toxicant (for example chlorine, Cl_2, hydrogen peroxide, H_2O_2) to selectively kill extended (and exposed) filamentous organisms while leaving the floc-forming organisms largely unaffected.

This chapter deals with the use of Cl_2 since our experience is with this chemical. Table 11.2 presents some information on concentrations of H_2O_2 found successful for bulking control.

Table 11.2 – Bulking Control with H_2O_2

Location	Reference	Successful concentration, mg/l
PETALUMA full-scale plant	ANON, FMC Corp. [2] Caropreso *et al.* [5]	9 – 68[a] average 31.5
ST. AUGUSTINE full-scale plant	ANON, FMC Corp. [3]	12 (basis unclear)
SAN JOSE Lab scale, fill and draw	Strünk, Shapiro [4]	40 – 200[a]
PRINCETON Small full-scale	Caropreso *et al.* [5]	100[b]
WILMINGTON Lab-scale	Cole *et al.* [6]	40 – 200[a]
WASHINGTON Pilot plant	Cole *et al.* [6]	20 – 40[b] 200 – 400[a]
TEXTILE MILL	Keller and Cole [7]	60 (basis unclear)

[a] Based on waste water inflow.

[b] Based on volume of aeration basin and clarifier (single dose).

11.3 CHLORINATION PARAMETERS

Chlorination has been used successfully for bulking control for many years [8–13]. A variety of dosing points and methods of determining Cl_2 dose have been used (Fig. 11.1 and Table 11.3). Without building additional mixing basins Cl_2 can be added at four locations:

(i) directly into the aeration basin;
(ii) into a sidestream pumping mixed liquor from, and returning it to, the aeration basin;
(iii) into the return activated sludge (RAS) stream;
(iv) into the flow from the aeration basin to the secondary clarifier.

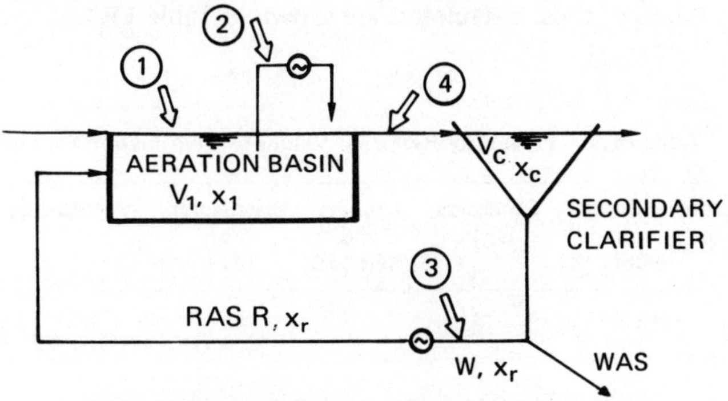

Fig. 11.1 – Chlorine dosing points for bulking control.

The Cl_2 dosing parameters are:

(a) Overall mass dose rate, T

$$T = \frac{M}{V_1x_1 + V_cx_c} = \text{g Cl}_2\text{/kg SS d}$$

(b) Dosed Cl_2 concentration at the Cl_2 dose point, C

$$C = \frac{M}{Q} = \text{mg Cl}_2\text{/l}$$

(c) Local mass dose rate at the Cl_2 dose point, T_m

$$T_m = \frac{M}{Qx_q} = \text{g Cl}_2\text{/kg SS}$$

(d) Exposure frequency of activated sludge to Cl_2 dosing, f

$$f = \frac{Qx_q}{V_1x_1 + V_cx_c} = \text{d}^{-1}$$

where:

M = Cl_2 application rate, g Cl_2/d
x = suspended solids concentration, g/l
V = volume, litres
Q = flow rate of stream into which chlorine is dosed, l/d

Subscripts refer to:

c = clarifier
1 = aeration basin
q = stream into which Cl_2 is applied.

Literature values of these parameters are shown in Table 11.3.

Table 11.3 – Previously Reported Values of Chlorination Parameters

Chlorination Parameter	Smith and Purdy [8]	Tapelshay [9]	Frenzel and Sarfert [10]	Eisenhauer *et al.* [11]	Jenkins [13]			Matsché [12]
$T = \frac{mgCl_2}{gSSd}$	–	–	–	2 – 7	5 – 15	4.5	5 – 11	–
$c = \frac{mgCl_2}{1}$	0.7 – 7.0	2.5 – 16.1	10.0 – 20.0	–	–	–	5 – 10	16
$f = d^{-1}$	–	–	–	–	–	–	–	–
$T_m = \frac{mgCl_2}{gSS}$		0.1 – 1.0	1.5 – 23	3.0 – 7.5		–		– –

T is independent of dosing point; C, T_m, and f can vary even though T remains constant, depending on Q and x values of the dosed stream. T and T_m are related by:

$$T_m = \frac{T}{f}$$

This makes it possible to describe chlorination for bulking control by three parameters, i.e. T or T_m, C and f.

There is not sufficient information to state conclusively the relative importance of the four chlorination parameters. Intuitively speaking. C should be important because it should control penetration of Cl_2 into flocs. Also, for Cl_2 to be effective the sludge inventory must be exposed to Cl_2 frequently enough to prevent the outgrowth of filamentous organisms. In practical applications, Cl_2 doses have often been expressed as T. Values in the range 2 to 15 g Cl_2/kg SS d have been found suitable for bulking control in systems with typical domestic treatment plant configurations (Table 11.3). Apparently this range of T values establishes suitable combinations of C and f at the Cl_2 dose point (usually RAS). Cl_2 application to long detention time systems based on typical domestic treatment plant T values gives very high C and very low f values. In such systems Cl_2 application to RAS may not control bulking. Figure 11.2 shows the application of 6 g Cl_2/kg SS d to a plant with a typical domestic waste configuration and to a long detention time system (for example for a high strength industrial waste). For the domestic waste case C = 7.5 mg/l and f = 6 d^{-1}, these values would be satisfactory for bulking control. For the industrial waste case, C = 121 mg/l and f = 0.37 d^{-1}; these values would not control bulking. The small amount of activated sludge that daily passed the dose point would be virtually destroyed.

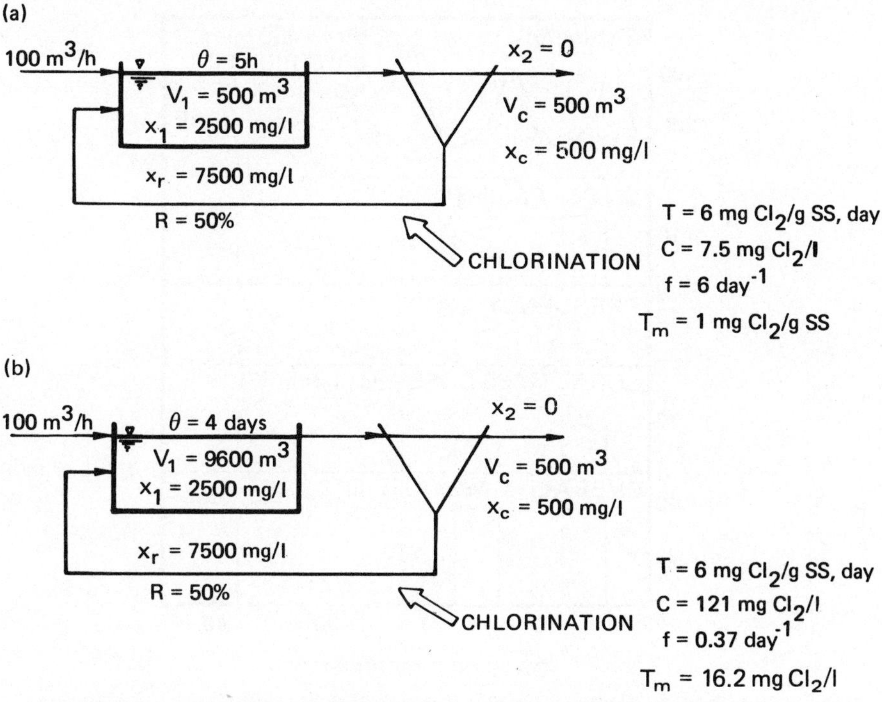

Fig. 11.2 – Chlorination parameters for a typical (a) domestic waste treatment plant and (b) high strength industrial waste treatment plant

Current research at SERL is seeking to determine critical values of chlorination parameters. Completely-mixed laboratory scale activated-sludge units are being operated on settled Richmond domestic sewage under conditions that promote severe bulking due to low concentrations of dissolved oxygen (DO) (COD loading = 0.6 kg/kg MLSS day; DO = 0.9 mg/l) [1]. The most abundant filamentous organism is Type 1701 [14]. Preliminary results (Fig. 11.3) indicate the effect of varying f at constant T, T_m, and C. For T = 36 g Cl_2/kg SS d, T_m = 6 g Cl_2/g SS and C = 12 mg Cl_2/l, f values of 6 d^{-1} and 4 d^{-1} cause SVI to decrease at about the same rate; when f is reduced to 3 d^{-1} the SVI decrease rate is significantly lower. Previous experiments under slightly different conditions (T = 12 g Cl_2/kg SS d, T_m = 6 –8 g Cl_2/g SS and C = 12 mgCl_2/l) indicated that f values of 2 d^{-1} and 3 d^{-1} did not control bulking.

C is expressed as dosed rather than residual Cl_2 concentration. RAS has a large Cl_2 demand. The type of Cl_2 residual (free or combined) may differ from one plant to another, most likely being strongly influenced by the ammonia and organic nitrogen content of RAS. Since disinfection rates of free and combined Cl_2 are different, the nature of the residual Cl_2 will affect its performance in bulking control.

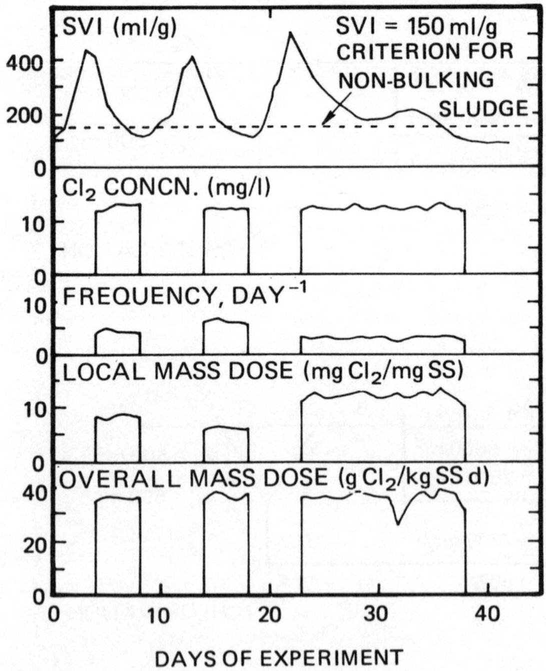

Fig. 11.3 – Effect of Cl_2 dosing parameters on rate of SVI reduction (laboratory activated sludge units treating settled domestic sewage).

11.4 EFFECT OF Cl_2 ON FILAMENTS AND FLOCS

The effect of Cl_2 on extended filamentous organisms has been assessed microscopically. Figure 11.4 shows the effect of Cl_2 on Type 1701 in laboratory experiments; Fig. 11.5 shows the effects of Cl_2 on *Thiothrix* sp in a full-scale plant. Both Type 1701 and *Thiothrix* sp are sheathed trichomes. Cl_2 successively causes cells to detach from the sheath and 'ball-up'; gaps then appear in the sheaths left by the disappearance of cells; finally, empty sheaths remain. With *Thiothrix* sp, this chain of events is preceded by the disappearance of intracellular sulphur granules (Fig. 11.5). Chlorination has successfully controlled bulking due to a variety of filamentous organisms in full-scale plants (Table 11.4). The relative susceptibility of various types of filamentous organism to Cl_2 is not known. This type of information will have to await the results of controlled experiments in which the Cl_2 residual (concentration and type) can be measured.

Table 11.4 – Bulking Filamentous Organisms That Have Been Successfully Controlled by Chlorination

Location	Major Filamentous Organism(s) Causing Bulking
City of Albany, GA	Type 0041 Type 0092 *Haliscomenobacter hydrossis* Type 1701
Malibu Mesa, CA	Type 1701 Type 021N
City of Los Angeles Terminal Island, CA	Type 021N Type 0581
Carrousel Pilot Plant, Brewery Waste	Type 0041 Type 0092
City of Pacifica, CA	*Sphaerotilus natans* Type 1701 Type 1863
City of San Jose, CA	*Haliscomenobacter hydrossis* Type 1701 Type 021N
Derry Township Municipal Authority, Hershey, PA	*Thiothrix sp.* *Microthrix parvicella* Type 021N
Oro Loma Sanitary District, CA	*Haliscomenobacter hydrossis* Type 1701

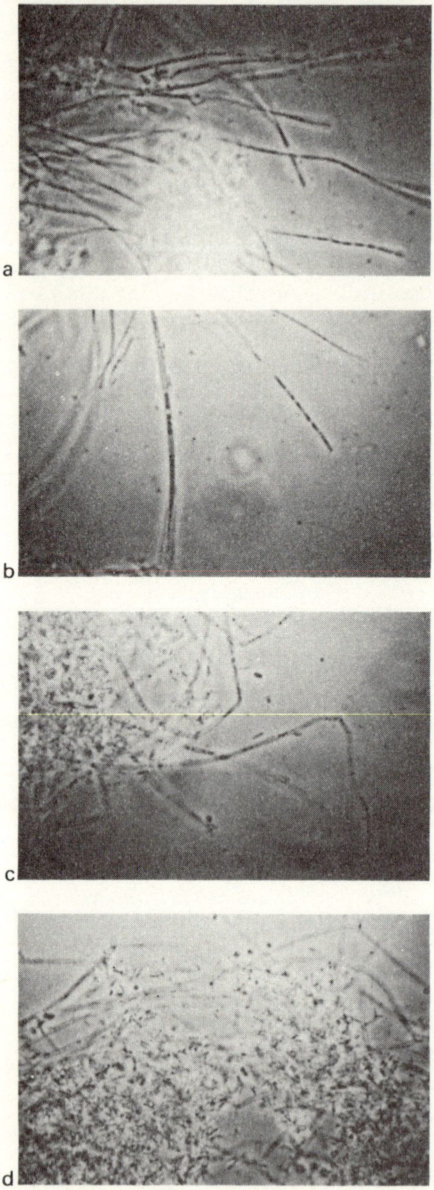

Fig. 11.4 – Effect of chlorination on Type 1701 in activated sludge (a) No chlorination (b) Moderate effects; cells 'ball-up' inside sheath, some empty space in sheath (c) Severe Cl_2 effects; many empty spaces in sheaths, few cells (d) Very severe Cl_2 effects; empty sheaths.

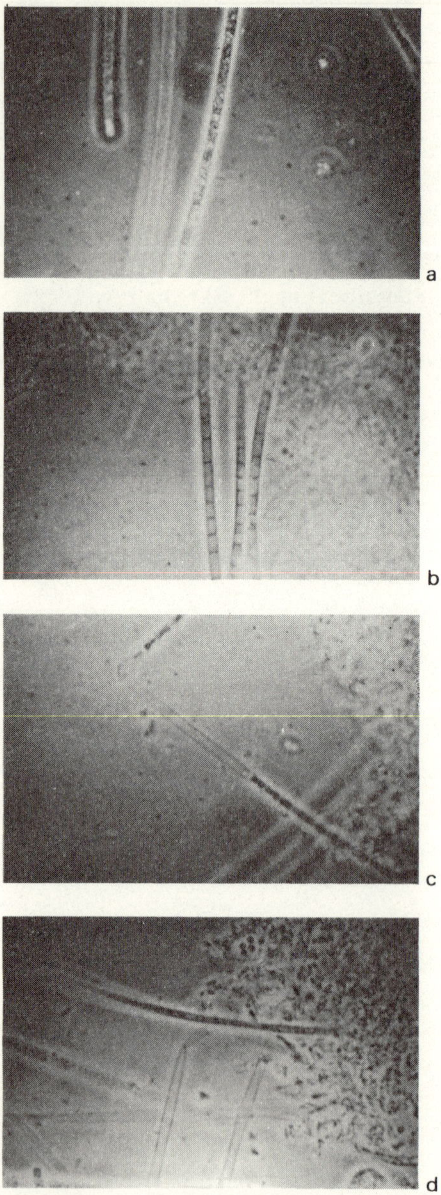

Fig. 11.5 – Effect of chlorination on *Thiothrix* sp in activated sludge (a) No chlorination; healthy *Thiothrix* sp containing sulphur granules (b) Moderate Cl_2 effects; no sulphur granules (c) Severe Cl_2 effects; cells 'ball-up' inside sheaths; empty spaces in sheaths (d) Very severe Cl_2 effects; empty sheaths.

Chlorine overdoses will eventually destroy the flocs. Overdoses can be detected by increases in effluent turbidity. When large, shock Cl_2 doses are applied to activated sludge, it has been reported that the effluent turns milky for about 24 hours, then clears up and the sludge settling improves markedly. The milkiness is due to large amounts of debris from floc destroyed by Cl_2. Figure 11.6 illustrates increases in secondary effluent turbidity with high overall Cl_2 mass doses in laboratory experiments. Table 11.5 presents data from the City of Los Angeles, CA, USA, Terminal Island Plant, which is a 115 000 m^3/d design activated-sludge plant receiving approximately 65 000 m^3 /d of sewage containing fish processing waste. The RAS Cl_2 injection point is a 100 mm diameter Cl_2 solution line tapped into the side of a 0.76 m diameter RAS pipe; inside the RAS pipe the Cl_2 solution line is reduced down to 25 mm diameter and extends to the pipe centre. Cl_2 solution is injected through a single port perpendicular to the RAS flow; RAS velocity is approximately 0.9 to 1.2 m/s; RAS travel time to the first aeration basin encountered is approximately 10 seconds. RAS chlorination was not practised during the entire month of July 1979. The high mean SVI value (300 ml/g) reflects lack of SVI control. High mean effluent SS values (42 mg/l) suggest that on occasions the sludge blanket overflowed into the effluent. During August and September 1979, RAS chlorination was practised – the average SVI was reduced to 120 ml/g in August 1979 and further to 50 ml/g in September 1979. During August 1979 secondary effluent SS averaged 9 mg/l and in September, 25 mg/l. When the average SVI increased to 63 ml/g (in October 1979) the mean effluent SS were at an intermediate value of 16 mg/l. These data support the field observations that activated sludge with very low SVI values (pin-point floc) produces effluents with poorer clarity than those from activated sludge with moderate (100 to 120 ml/g) SVI values. In this instance over-chlorination converted a bulking sludge into a pin-point floc.

Table 11.5 – SVI and Secondary Effluent SS Levels During Use of RAS Chlorination: City of Los Angeles, Terminal Island, California

Date, 1979	Sludge Volume Index (ml/g)			Secondary Effluent SS, (mg/l)		
	High	Low	Mean	High	Low	Mean
Jul	519	176	302	152	2	42
Aug	180	39	120	17	3	9
Sep	86	36	50	57	11	25
Oct	84	42	63	39	9	16

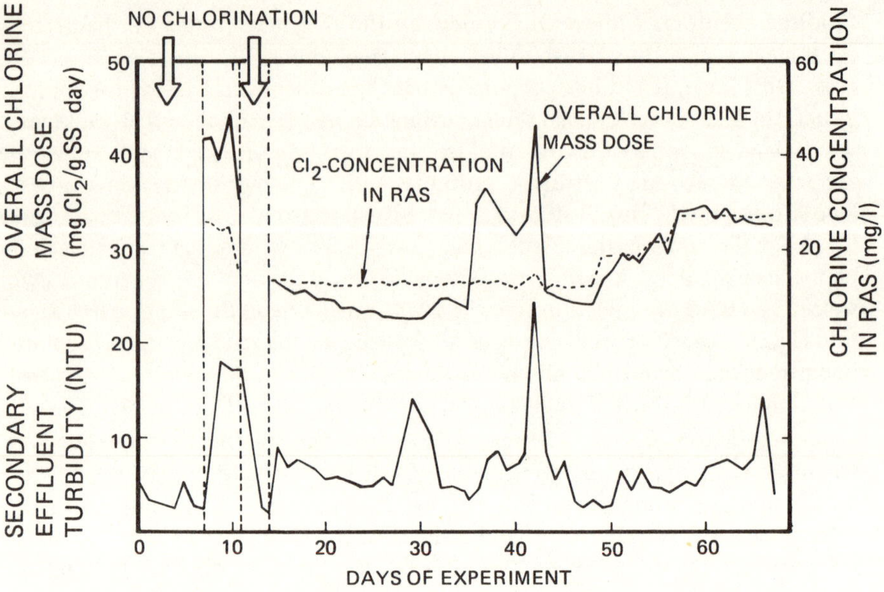

Fig. 11.6 – Effect of RAS chlorination on secondary effluent turbidity (laboratory activated-sludge units).

11.5 PRACTICAL APPLICATION OF CHLORINATION FOR BULKING CONTROL

Several guidelines should be followed to ensure successful chlorination for bulking control:

(a) A target SVI (or some other measurement of activated-sludge settling) should be established

The target SVI should be the value below which the plant can be operated satisfactorily and economically without any of the problems associated with poorly settling activated sludge. In our opinion it is best to select the highest target SVI possible, consistent with satisfactory and economic operation. Activated sludges with high SVIs (e.g. $\geqslant$ 150 ml/g) tend to produce clearer (low turbidity) effluents, lower in SS and BOD_5 than those from activated sludges with low SVIs (e.g. 50 to 100 ml/g).

(b) Chlorination should only be used when the target activated sludge settling property is significantly and consistently exceeded

A plot of daily (or more frequent) SVI values should be made so that it can be judged when the target SVI is being approached and when it is exceeded significantly and consistently. Examination of this plot also allows adjustment of Cl_2 dose to anticipate SVI changes.

(c) Cl_2 should be added in known and controlled doses to all of the activated sludge at a point of excellent mixing

(d) Cl_2 should be added at a point where Cl_2 demand from the waste is at a minimum

(e) Cl_2 should be dosed on the basis of the sludge inventory of the plant (kg Cl_2/1000 kg SS d)

Where the sludge inventory in the secondary clarifiers is significant, this must be included in calculating total solids inventory. For 'two basin' processes (for example sludge reaeration, contact stabilization), the entire sludge inventory of both basins should be determined.

(f) Cl_2 concentration (mg/l) at the dosing point and the number of exposures of the solids inventory to the Cl_2 dose should be considered in choosing a dosing point

The magnitude of the Cl_2 dose required for bulking control is not well defined – possibly because many factors such as mixing conditions, and Cl_2 demand vary from one situation to another. Some approximate practical guidelines are:

(i) For nitrifying plants with typical municipal configuration, the maximum Cl_2 dose should not exceed 5 kg Cl_2/1000 kg SS d.
(ii) If RAS chlorination is being initiated at a treatment plant for the first time and there is not a 'crisis situation' in which bulking must be cured immediately it is good practice to start at low Cl_2 doses (for example 2 kg Cl_2/1000 kg SS d). If this dose rate, applied for several days, does not lower or stabilize SVI, the dose should be increased slowly. If doses of much greater than 9 to 10 kg Cl_2/1000 kg SS d are required, then causes should be sought (for example unusually high Cl_2 demand, poor mixing and inaccurate Cl_2 dose measurement).

When a given Cl_2 dose begins to achieve stabilization and then the reduction of SVI, the dose should be decreased slowly. The objective is that as the target SVI is approached, the Cl_2 dose should be zero. The Cl_2 should not be turned on again until the target SVI is significantly and consistently exceeded.

Guideline *(c)* can be best met by injecting Cl_2 solution into RAS flowing in a common closed pipe, preferably at a point of turbulence induced by the RAS pumps or by a piping elbow.

Ideally, a separate chlorinator should be dedicated to RAS chlorination because the Cl_2 dose rates required for bulking control usually are much lower than those employed for secondary effluent chlorination. However, since it is not common current practice to design RAS chlorination with a separate dedicated chlorinator, a Cl_2 solution line can be taken off an effluent chlorinator to provide Cl_2 solution for bulking control. It is imperative to know accurately the Cl_2 solution flow in this line and an independent rotameter is mandatory.

If there is no common RAS line, chlorination devices must be installed for each RAS line. Individual control of Cl_2 solution to each point should be available. Excellent mixing of Cl_2 solution and RAS is of the utmost importance. The Cl_2 demand in RAS is so rapid that, without adequate mixing, a large part of the RAS will not be contacted by the Cl_2 while a small part will be overdosed. A further implication of the rapid Cl_2 demand in RAS is that almost no contact time between Cl_2 and RAS is required. One successful application of RAS chlorination that illustrates this point is at the City of Pacifica, CA, USA. At this 16000 m^3/d plant the RAS system is by travelling bridges each of which have five air-lift tubes activated by an air-lift pump. The tubes discharge to a common channel and then directly into the aeration basin. The chlorination system (Fig. 11.7) is a 70 kg Cl_2 gas cylinder and a Wallace and Tiernan V–100 Chlorinator mounted over the travelling bridge discharge. Water for Cl_2 solution is pumped from near the secondary effluent weirs. The Cl_2 solution passes through a 38 mm PVC pipe to a PVC manifold placed in the RAS discharge to the aeration basin. This horizontal plastic pipe has 3 mm drilled holes spaced approximately 25 mm apart which face in a direction opposite to the RAS flow. Excellent mixing is obtained with this arrangement. The contact time before the RAS enters the aeration basin is certainly less than 1 second.

(g) Reliable sludge settling and effluent quality measurements must be made
Chlorination for bulking control entails the purposeful addition of a toxicant to a biological system. Because of this, control tests to assess effects of Cl_2 addition on settling properties and effluent quality must be performed frequently and carefully.

Case History: City of Albany, GA, USA
The City of Albany, GA, USA, activated-sludge plant is a 76 000 m^3/d (design) currently receiving approximately 54 000 m^3/d of a waste in which approximately 50 per cent of the BOD_5 and SS loads are from industry (paper processing and convenience foods manufacturing). The plant experienced bulking problems to such a degree that with only an approximate 45 000 m^3/d influent flow, attainment of secondary effluent criteria (30 mg/l) for BOD_5 and SS was not possible.

At Albany, the permanent RAS chlorination system injects Cl_2 solution into a series of well-mixed RAS wet wells. The installation consists of a Wallace and Tiernan Chlorinator (1800 kg/d capacity), and a 75 mm injector with 75 mm PVC injector water and Cl_2 solution lines. The Cl_2 injection system is enclosed in a sheet metal housing for protection. The Cl_2 solution line is a 75 mm PVC line leading to a 4.5 m long, 75 mm PVC header with holes drilled in it every 150 mm ranging in size from 9.5 mm to 16 mm diameter. The system was constructed in November 1977 for $ 4953.

Fig. 11.7 – RAS chlorination system on travelling bridge at Pacifica (a) Travelling bridge RAS chlorination (b) Cl_2 cylinder for Cl_2 solution (c) Cl_2 injection point in RAS on travelling bridge.

When chlorination for bulking control was initiated, the RAS chlorination system did not exist. Because of the urgency to control SVI, Cl_2 was added to a wet well where the entire RAS stream mixed with primary effluent. This mode of chlorination was commenced on Week 15 (Fig. 11.8) and continued until Week 26 when the change to RAS chlorination was made. When Cl_2 was being added to the mixture of RAS and primary effluent, doses of 5 to 15 kg Cl_2/1000 kg SS d were needed to control bulking. Immediately after the change from chlorinating the mixture of RAS and primary effluent to chlorinating RAS alone, Cl_2 doses (7 kg Cl_2/1000 kg SS d) that satisfactorily controlled bulking for the mixture of RAS and primary effluent were 'overdoses' for RAS alone. Figure 11.8 shows that even though the Cl_2 dose declined during changeover of Cl_2 dose points (from 4.7 kg Cl_2/1000 kg SS d to 2.9 kg Cl_2/1000 kg SS d) there was an increase in secondary effluent SS and (to a lesser extent) in secondary effluent BOD_5.

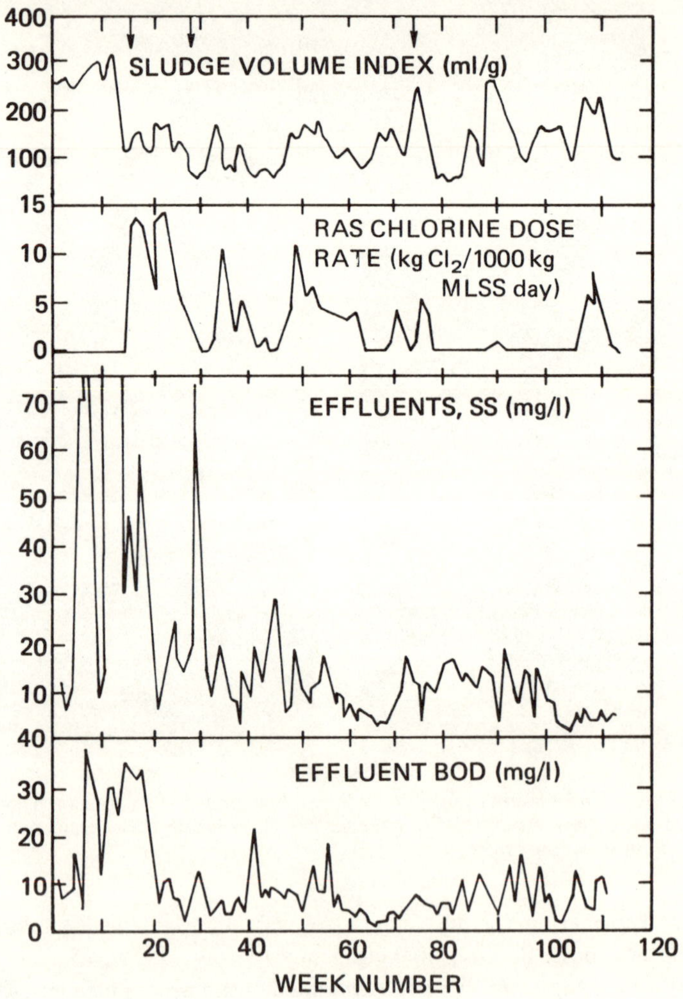

Fig. 11.8 – Operating data, City of Albany, GA., waste water treatment plant.

During the period covered by Fig. 11.8 changes were made in target SVI. For Weeks 27 to 71, the target SVI was 100 to 150 ml/g; for Weeks 72 to 122, the target SVI was 200 to 250 ml/g. Increasing target SVI has had several beneficial effects. These are:

- Cl_2 usage dropped from an average of 2.9 kg/1000 kg SS d at a target SVI of 100 to 150 ml/g, to 0.86 kg/1000 kg SS d at a target SVI of 200 to 250 ml/g (Table 11.6).
- Effluent BOD_5 and SS values have decreased as the target SVI has been increased. At a target SVI of 100 to 150 ml/g, secondary effluent values were 7 mg BOD/l and 12 mg SS/l; at a target SVI of 200 to 250 ml/g secondary effluent values were 7 mg BOD/l and 8 mg SS/l, (Table 11.6).

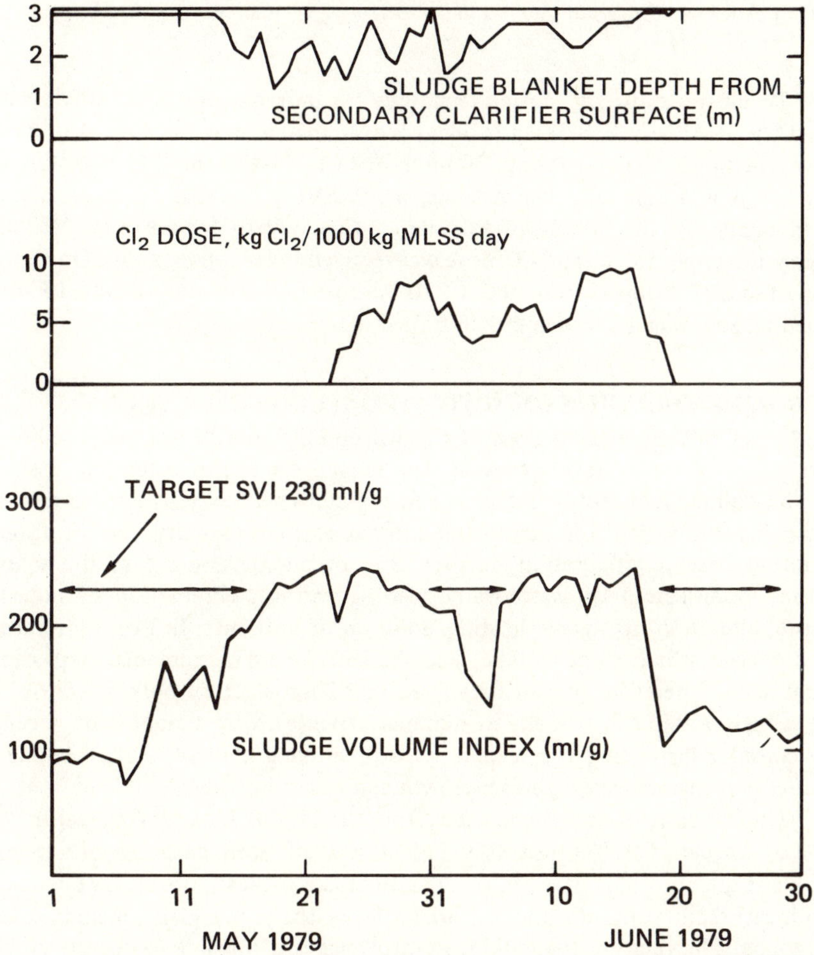

Fig. 11.9 – RAS chlorination on dose basis with target SVI for bulking control.

Table 11.6 – RAS Chlorine Usage and Effluent Quality at Albany, GA

Period Week No.	Bulking Control Method	Target SVI Range (ml/g)	Average SVI During Period (ml/g)	Average Chlorine Dose for Bulking Control (kg Cl_2/1000 kg MLSS Day)	Secondary Effluent BOD_5 (mg/l)	Secondary Effluent SS (mg/l)
0 – 14	Prior to Chlorination	–	254	0	22	71
15 – 26	Chlorination of RAS, Primary Effluent Mixture	100 – 150	138	10.8	18	23
27 – 71	RAS Chlorination	100 – 150	116	2.9	7	12
72 – 122	RAS Chlorination	200 – 250	161	0.86	7	8

To illustrate the Cl_2 dosing technique for bulking control, as employed by the City of Albany, a bulking incident is examined in detail in Fig. 11.9. At this time, the target SVI was about 230 ml/g. The Cl_2 dose to the RAS was started at low levels and gradually increased until the SVI responded by stabilizing and then falling. In the middle of this period (30 May to 6 June) the SVI fell to below the target value. The Cl_2 dose was reduced in response to this. On 18 to 20 June the SVI dropped from about 230 ml/g to approximately 100 to 130 ml/g. The Cl_2 dose was reduced and then turned off.

11.6 SOME THOUGHTS ON ACTIVATED SLUDGE OPERATION

Earlier in this chapter, the use of Cl_2 for bulking control was proposed as the first stage of a two-stage approach. The second stage is to determine the cause of the bulking and rectify the problem. The steps necessary to achieve this stage are many and varied, for example additional aeration capacity, modification of aeration basin configuration, introduction of anoxic zones into the aeration basin, modification of waste-water feeding pattern, elimination of industrial waste, alteration of organic loading, addition of nutrients. In many instances it is not clear which steps to take; also the cost for taking remedial steps can be significant. Therefore, it can be argued that if satisfactory bulking control can be achieved with Cl_2 (usually at minimal expense), why should it be necessary to consider instituting the second stage of bulking control – i.e., seeking the cause(s) of filamentous organism growth and rectifying them?

On the basis of data such as that from the City of Albany, GA, and the City of Los Angeles, CA, Terminal Island Plant, it would seem advantageous, from an effluent quality point of view, to operate at the highest possible SVI consistent with satisfactory and economical operation of the entire plant, including solids handling. Chlorination for bulking control makes it possible to operate at whatever SVI is desired. It is no longer necessary to consider SVI values of approximately 100 ml/g as normal. The practice of operating at about this SVI value appears to have been adopted largely because SVI values significantly higher than this indicated that unstable operation was imminent, i.e. that the SVI would probably be much higher the next day and soon out of control. Chlorination using the techniques outlined in this chapter can allow stable operation at any SVI value.

Another interesting possiblity presented by the ability to control SVI at any value concerns energy saving in activated-sludge plants. Efficient BOD removal can be achieved at aeration basin DO values much lower than those required to prevent the production of sludge bulking due to low DO concentration. Often, minimum DO levels of 2 mg/l in air-aerated, non-nitrifying activated-sludge plants are maintained for control of bulking. Significant energy savings should accrue if plants were operated at very low aeration basin DO levels (for example 0.1 to 0.5 mg/l DO in the effluent) and any bulking caused by this mode of operation were to be controlled by chlorination. Since it may be possible to operate an activated-sludge plant under these conditions at a higher than normal SVI, the dual benefits of improved effluent quality (low SS) and energy savings through lower air requirements may be obtained.

11.7 CONCLUSIONS

- A two-stage approach to bulking control is proposed and has been found successful in many activated-sludge plants. The first stage is chlorination of activated sludge at a well-mixed point that provides sufficiently frequent exposure of the activated-sludge inventory to the Cl_2 and using a target SVI value to determine the dose.
- The preferred dosing point is usually the RAS; however, in plants with a large solids inventory (long detention time) dosing of Cl_2 to the mainstream of the plant may be necessary.
- Chlorination is effective for controlling a wide variety of filamentous organisms.
- Activated sludges with high SVI values generally produce effluents of excellent clarity and low BOD and SS concentrations. Chlorination allows stable SVIs to be maintained at any SVI value.
- Establishment of conditions in an aeration basin (for example low DO) that cause filamentous outgrowth coupled with controlling SVI to a desired stable, moderately high value by chlorination is suggested as a technique for producing effluents approaching tertiary treatment quality (10 mg SS/l and 10 mg BOD_5/l) together with attaining operating economies through reduction in power consumption for air supply.

11.8 ACKNOWLEDGEMENTS

There are many people to whom we owe a debt of gratitude for providing information and conducting work that went into preparing this chapter. Dr. Peter F. Strom, Post Doctoral Research Fellow, Sanitary Engineering Research Laboratory, University of California, Berkeley, performed some of the filamentous organism identifications. Data on plant operation were provided by Mr. Russ Edwards, Assistant Director of Public Works, City of Albany, GA, USA; Mr. Kenneth L. Ludwig, City of Los Angeles, CA, USA; Mr. William Wilkens, REMCON Inc., Burlingame, CA, USA; Mr. Robert E. Beebe, San Jose – Santa Clara Water Pollution Control Plant, San Jose, CA, USA. Some of the work was conducted with support from the US National Science Foundation.

11.9 REFERENCES

[1] Palm, J. H., Jenkins, D. and Parker, D. S. The relationship between organic loading, dissolved oxygen concentration and sludge settleability in the completely mixed activated sludge process. *Journal of the Water Pollution Control Federation,* 1980, **52**, 2484-2506.

[2] Anon. FMC Corporation, Bulking control with hydrogen peroxide: case history, City of Petaluma water pollution control plant, Sonoma County, CA. *Technical Data Pollution Control Release No. 41,* 1973.

[3] Anon. FMC Corporation, Pollution Control Release No. 95, Sludge bulking cure: hydrogen peroxide. *Municipal South,* November/December, 1976.

[4] Strünk, W. G. and Shapiro, J. Bulking control made easy – with hydrogen peroxide. *Water and Pollution Control,* 1976, **11**, 10–14 and 40–41.

[5] Caropreso, F. E., Raleigh, C. W. and Warner, J. C. Attack bulking sludge with H_2O_2 and a microscope. *Bulletin, California Water Pollution Control Association,* 1974, No. 44.

[6] Cole, C. A., Stamberg, J. B. and Bishop, D. F. Hydrogen peroxide cures filamentous growth in activated sludge. *Journal of the Water Pollution Control Federation,* 1973, **45**, 829–836.

[7] Keller, P. J. and Cole, C. A. H_2O_2 controls bulking. *Water and Wastes Engineering,* 1973, **10**, E4–E7.

[8] Smith, R. S. and Purdy, W. C. Use of chlorine for the correction of sludge bulking in the activated sludge process. *Sewage Works Journal,* 1936, **8**, 223-230.

[9] Tapelshay, J. A. Control of sludge index by chlorination of activated sludge. *Sewage Works Journal,* 1945, **17**, 1210–1226.

[10] Frenzel, H. J. and Sarfert, F. Gefahrungen und die Verhinderung von Blahschlammbidung durch Chlorung der Belebten Schlammes. *Gas und Wasserfach, Wasser, Abwasser,* 1971, **112**, 604–606.

[11] Eisenhauer, D. L., Sieger, R. B. and Parker, D. S. Design of integrated approach to nutrient removal. *Journal of the Environmental Engineering Division ASCE,* 1976, **102**, 37-54.

[12] Matsche, N. Blähschlamm – Ursachen und Bekämfung. Wiener Mitteilingen. *Wasser – Abwasser – Gewasser,* 1977, 22, 1.

[13] Jenkins, D. The control of activated sludge bulking. Presented at the *52nd Annual Conference, California Water Pollution Control Association,* 1980, Monterey, California.

[14] Eikelboom, D. H. Identification of filamentous organisms in bulking activated sludge. *Progress in Water Technology,* 1977, **8**, (6), 153-162.

Discussion

Chairman: **Professor P. Grau** (Institute of Chemical Technology, Prague, Czechoslovakia)

Mr. A. G. Boon (Water Research Centre) opened the discussion by saying that he had been present at a previous conference several years ago when Professor Jenkins had presented a paper describing how bulking could be controlled by the use of high concentrations of dissolved oxygen. Professor Jenkins' work had shown that *Sphaerotilus natans* could be effectively controlled by this method but later work and many of the papers at this conference (chapters in this book) had indicated that *Sphaerotilus natans* was not a particularly important bulking organism. Mr. Boon asked if Professor Jenkins agreed with this statement and also asked if other filamentous micro-organisms could be controlled by the use of appropriate dissolved-oxygen concentrations.

Mr. Boon continued by referring to Professor Jenkins' latest work on chlorination as a control method for activated-sludge bulking and asked if Professor Jenkins considered chlorination to be effective against all types of filamentous micro-organisms, or whether it would only be useful in certain circumstances. **Professor Jenkins** replied that he was not prepared to state that the use of chlorine would cure bulking in every instance, but it had proved successful in all the cases he had investigated so far. He went on to say that the frequency of exposure of the micro-organisms to the chlorine at the dosing point was an important criterion and if this frequency was high enough, successful results would be obtained. Chlorine dosing would fail to cure bulking in circumstances where only a small fraction of the total activated-sludge inventory was exposed to a high concentration of chlorine. Professor Jenkins said that his paper (Chapter 11 of this Book) contained a list of filamentous micro-organisms that had been present in the activated sludges at the plants where chlorine dosing had been applied.

Referring to his previous work on the relationship between sludge loading, dissolved-oxygen concentration and sludge settleability in completely-mixed systems, Professor Jenkins said that he stood by his results and believed that *Sphaerotilus natans* and Type 1701 could be controlled by manipulating the concentration of dissolved oxygen. In plants where the sludge loading rate was low, such as extended aeration systems, other filamentous micro-organisms, such as *Microthrix parvicella,* Type 021N, Type 0092 and Type 0041, were often present and he did not think that the dissolved-oxygen concentration was important in eliminating such species. However, they could be successfully controlled by the use of chlorine.

Professor Jenkins said that the rapid increase in the cost of energy in recent years meant that the use of high dissolved-oxygen concentrations to control sludge settleability could not always be justified from economic considerations. Indeed he thought that it was cheaper, in many cases, to operate a treatment plant under conditions of low dissolved oxygen and control any resulting bulking problems by the use of chlorine. This strategy is particularly attractive in the USA where it is standard practice to install chlorination equipment on site for treatment of final effluent before discharge. The chlorine dosage required for control of sludge settleability would be very small in comparison to that required for effluent disinfection.

Dr. K. C. Thompson (Yorkshire Water Authority, UK) said that disposal of sewage sludges to land was very common in the UK and he asked if there was any risk of chlorinated compounds being absorbed into such sludges if chlorination was practised at a treatment works. **Professor Jenkins** replied that he did not consider that there was any health risk resulting from the formation of chlorinated compounds. Chlorination doses employed were very small and most of the dose would react almost instantaneously with ammonia and micro-organisms, leaving only a minute fraction to produce chlorinated organic materials. Professor Jenkins was of the opinion that there were many substances in sewage sludges which were very much more important toxicologically than the chlorinated compounds under discussion. His personal view was that disposal at sea was preferable to land disposal of sludge since it provided a less direct route back to the human population.

Dr. Thompson agreed with Professor Jenkins' comments on the amount of chlorinated compounds that might be produced in sludges as a result of chlorine dosing and continued by saying that the health risks associated with land disposal of sludges were an emotive issue and hence the effects of all new treatment techniques on this aspect needed careful consideration.

Professor Jenkins said that he had work in progress which would address this point and he had already concluded that chlorine dosing might not be suitable at plants treating certain industrial wastes. He did not think, for example, that it would be wise to apply chlorination techniques to a plant treating petrochemical waste since such wastes contain high proportions of organic materials which would be readily chlorinated. In the USA wastes produced at petrochemical and refinery complexes are treated separately from domestic wastes generated at such complexes to avoid chlorination. Professor Jenkins added that hydrogen peroxide might be more suitable to use as a chemical treatment method for control of bulking in plants treating petro-chemical wastes.

Dr. C. F. Forster (University of Birmingham, UK) said that the chlorination of final effluents from treatment works would result in chlorine being added to nitrogen compounds which had been oxidized. However, he considered that Professor Jenkins' proposal involved the addition of chlorine to raw sewage and hence there was a potential for the formation of chloramines, which might be ecologically hazardous.

Professor Jenkins said that the requirement to discharge oxidized nitrogen in final effluents was by no means universal and many plants in the USA produced effluents containing ammonia. He went on to say that his proposals did not, in fact, involve the chlorination of raw sewage since the chlorine was added at a point of good mixing in the recycled activated sludge. Professor Jenkins said it was not possible to measure residual free chlorine in the flow of recycled sludge entering aeration tanks where chlorination was practised.

Mr. J. Arundel (Thames Water Authority, UK) asked Professor Jenkins if he could provide some information on the cost of chlorination as a method of bulking sludge control. **Professor Jenkins** replied that the maximum dose of chlorine required would be about 10 kg of chlorine for every tonne of activated-sludge solids in the system per day. In the USA the cost of chlorine was about 38 cents/kg and he understood it was about 27 pence per kg in the UK. Hence for a system containing 100 000 kg of activated sludge the cost would be about £27 per day for chlorine. Professor Jenkins considered that this was a relatively low figure in comparison to the degree of control which it accomplished.

Dr. J. Houtmeyers (Forest Biology Centre, Belgium) referred to a statement Professor Jenkins had made earlier in the discussion, namely that it was cheaper to chlorinate the recycled sludge to control poor settleability than to extend the aeration capacity. Dr. Houtmeyers speculated that, in the future, treatment plants might be built in two stages. The first stage being operated at a high loading until a time when a higher quality effluent was required when the second stage could be added, resulting in a lower loading rate overall. Dr. Houtmeyers considered that Professor Jenkins' recycled sludge chlorination techniques might be useful in improving and controlling any poor sludge settleability resulting from the operation of the initial, highly loaded plant.

Professor Jenkins agreed with Dr. Houtmeyers' philosophy and added that he would not design an activated-sludge plant without facilities for the chlorination of recycled sludge.

Miss A. J. Blachford (Water Research Centre) said that Professor Jenkins chlorination methods were based essentially on continuous dosing, whereas when chlorine dosing had been employed in the UK to control bulking, a shock dose had usually been applied. The theory of shock dosing being that if all the filaments were eliminated from the system then they should not develop again in the future. She referred to one of Professor Jenkins' figures which indicated that the filaments re-appeared when the continuous dosing was stopped and asked if this may not have been the case if the dosing had been continued for a longer period. Miss Blachford also asked if chlorine dosing reduced the treatment capacity of a works due to its toxic effect. **Professor Jenkins** replied that if the ecological conditions in an activated-sludge plant favoured the growth of filamentous species, then such organisms would return after chlorine dosing was stopped. He gave the example of *Sphaerotilus natans* developing in an aeration tank operated at a low dissolved-oxygen concentration and then being controlled by the use of chlorine. He said that Sphaerotilus would redevelop in the system when chlorine dosing was stopped, if the dissolved-oxygen concentration was maintained at a low level.

Professor Jenkins said that he preferred to think of his chlorine dosing techniques as 'fine tuning' rather than intermittent chlorination. He was not in favour of shock-dosing methods which in the past could mean the dumping of truck loads of calcium hypochlorite into the aeration tanks. Such methods usually resulted in the temporary elimination of all the activated-sludge microflora.

In answer to Miss Blachford's second question, Professor Jenkins referred to a statement of Mr. Eikelboom's when he had indicated that the structure of activated-sludge flocs resulted in only about one ten thousandth of the microbial mass being externally exposed. Thus, most of the biomass was inside the floc and would not be affected by chlorination, providing the system was not overdosed. Professor Jenkins did not consider that the removal of such a small fraction of the biomass would result in a noticeable reduction in treatment capacity.

Mr. D. H. Eikelboom (Research Institute for Environmental Hygiene, TNO, The Netherlands) referred to one of Professor Jenkins' figures which showed that the BOD of a final effluent during chlorine dosing was about 10 mg/l. Mr. Eikelboom's experience with chlorine addition in Holland was that the effluent BOD usually increased to 20–30 mg/l. **Professor Jenkins** said that such effects had never been observed in his work and he asked if Mr. Eikelboom was referring to chlorine additions to lowly-loaded oxidation ditches. **Mr. Eikelboom** replied that the plant in question was highly loaded and added that the chlorine dose employed had been about 2.5 mg chlorine per g dry solids. The dose was added to the recycled sludge. Mr. Eikelboom said that the deterioration in effluent quality was associated with a decrease in Protozoa. **Professor Jenkins** agreed that chlorine could cause a reduction in the protozoan population.

Dr. J. Cuthbert (Water Research Centre) said that he appreciated that Professor Jenkins' technique involved intermittent dosing of chlorine to keep the SVI value below some predetermined target value which would vary from plant to plant, but he asked if Professor Jenkins could give some indication of the fraction of time for which chlorine would be dosed at a typical plant. **Professor**

Jenkins presented a slide which referred to chlorine treatment at a plant in Albany, Georgia, USA over a period of 110 weeks. He said that the plant suffered severe settleability problems due to inadequate dissolved-oxygen concentration and the treatment of effluent from paper and confectionery manufacture. The chlorine dosage was required for about two-thirds of the period. Professor Jenkins emphasized that the example referred to a particularly difficult case and that chlorine usage might be considerably less at other plants.

Chapter 12

Some experiences in the control of bulking activated sludge

C. B. Waller and **B. J. E. Hurley**, Thames Water Authority

12.1 INTRODUCTION

The bulking of activated sludge is generally recognized as a condition in which the sludge exhibits poor settlement characteristics because of loose, flocculent, cotton-wool like growths of filamentous micro-organisms [1,2].

The factors which can encourage this phenomenon are numerous and include over-loading, variations in flow regime, under-aeration, over-aeration, shock loads, and high C:N and C:P ratios. Certainly a 'completely-mixed' surface aeration system receiving a waste high in carbohydrate content might be expected to experience bulking. This was the situation encountered at a Lambourn Division Sewage Works in 1975 when the Scientific Services Directorate of the Thames Water Authority was requested to investigate and advise upon methods which might be employed to correct the problem at this works.

12.2 THE PROBLEM AT THE LAMBOURN DIVISION SEWAGE WORKS

Prior to 1975 the activated-sludge plant suffered from severe bulking problems which culminated at times in loss of activated-sludge solids to the receiving stream. This occurred particularly during the summer months due to the inclusion of up to 25 per cent by volume of a trade discharge rich in carbohydrate material from a fruit and vegetable canning factory. The problem was most pronounced during the months June to August when potatoes were being processed and the concentration of glucose in the trade waste approached 200 mg/l.

During the potato seasons of 1972 and 1973 the problem was partially alleviated by diverting a portion of the flow from the new works (built in 1972), to the old biological filter works. Amelioration by this method became impossible when a connecting sewer between the works was sealed. This situation led to the building of extensions to the new works which were completed in time for the 1974 season.

During the latter part of the 1974 potato season, commercial oxygen was used to supplement the overloaded aeration system at times of peak loading. Whilst some improvements were achieved, the sludge bulking persisted and it became necessary to initiate objective investigations to contend with the problems in the longer term.

These investigations were aimed at both:

(i) initially containing bulking and preventing the discharge of activated-sludge solids to the receiving watercourse,

and

(ii) providing a more permanent solution to the bulking problem.

An initial assessment of the situation indicated some deficiencies in the hydraulic flow distribution throughout the works and measures were taken to correct this. The aeration lanes, which were fitted with surface aerators, were found to be 'completely-mixed' with respect to suspended solids, ammonia and dissolved-oxygen levels.

Analysis of the settled sewage indicated that during periods of high trade waste load a nutrient deficiency may have existed, but taken over a 24-hour period nitrogen and phosphorus were 'non-limiting' and therefore not the major cause of the sludge bulking.

Average analyses of 24–hour composite flow proportioned samples of the trade waste and crude and settled sewage are shown in Table 12.1.

Table 12.1 – Analysis of Trade Waste, Raw and Settled Didcot Sewage

	pH-value	Suspended solids (mg/l)	BOD_5 (mg/l)	COD (mg/l)	NH_3–N (mg/l)	Total P (mg/l)
Trade Waste	5.5	260	1360	1920	4.7	19.6
Didcot Raw Sewage	6.8	320	560	1110	21.8	33.4
Didcot Settled Sewage	6.9	200	340	610	22.9	29.8

12.3 METHODS USED FOR CONTROL OF THE BULKING PROBLEM

The following measures were used in an attempt to contain the bulking sludge and prevent the loss of activated-sludge solids to the receiving water course.

(i) Attempted reconstitution of the sludge
An attempt was made to reconstitute the sludge by replacing the bulking sludge, after large-scale surplussing, with a high quality sludge obtained from a nearby works. The new sludge rapidly deteriorated with filamentous bacteria again becoming the dominant organisms present.

(ii) Flow balancing
Variations in diurnal flow were minimized by diverting the crude sewage to two storm tanks in order to maintain the maximum rate of flow at about 5 000 m^3/day. This helped to maintain the sludge blanket level in the final settlement tanks at a minimum depth of about 0.3 m. At times when sludge settlement was particularly poor all crude sewage flow was diverted to the storm tanks. When these became full it was necessary to pump either settled sewage or final effluent to two areas of land within the curtilage of the works to prevent the discharge of solids to the stream.

(iii) Lime treatment
An early attempt was made to reduce the organic load on the activated-sludge unit by the addition of lime upstream of primary sedimentation. Powdered lime was added to the crude sewage wet well at a rate (~80 kg/h) sufficient to obtain a pH-value of 9 to 10. The effect of liming was difficult to assess as analyses of the limed and unlimed settled sewage were inconsistent. No significant improvement was indicated by employing this method of treatment.

Large quantities of primary sludge were produced at this time which resulted in frequent de-sludging of the primary sedimentation tanks. However, no detrimental effect on the gas production of the anaerobic digestion plant was observed.

It was also found, both in laboratory and on full-scale plant experiments, that liming of the mixed liquor did not assist in the settling of activated sludge in the final tanks.

(iv) The use of commercial oxygen
The residual dissolved oxygen in certain aeration lanes became very low at periods of high loading and side stream oxygen injection using the BOC Vitox system was implemented. Some improvement of DO was achieved but at times of high load the oxygen uptake was still greater than the input supplied by both the surface aerators and Vitox equipment.

(v) Coagulation using a polyelectrolyte
The previously-mentioned measures were helpful in limiting the bulking problem, but it became obvious that some type of flocculation aid was required to assist sludge settlement in the activated-sludge plant final settlement stage. Lime proved unsuccessful and cationic polyelectrolyte was considered. Full-scale dosing of mixed liquor with a polyelectrolyte is not standard sewage works practice and

little information is available on its effect on plant performance or the subsequent effect of the final effluent on the watercourse to which it is discharged. Consequently, it was decided to keep the dose rate to the absolute minimum necessary to effect flocculation and, at the same time, note any detrimental effects on the plant or receiving stream.

Zetag 94 was chosen as the preferred polyelectrolyte, and laboratory tests showed that the minimum effective dose was 6 mg/l, (corresponding to 0.05 per cent active matter to solids). A 10 per cent solution of Zetag 94 was made up in 0.45 m^3 tanks and dosed into the inlet of each final tank by means of a peristaltic pump. No adverse effects on the plant or the stream were observed either at this concentration or when it became necessary to increase the dose rate to 12 mg/l (0.1 per cent active matter to solids). This concentration of polyelectrolyte improved the Sludge Volume Index (SVI) of the mixed liquor from 872 to 170 ml/g (not corrected for suspended solids concentration) and the Stirred Specific Volume Index (SSVI) at 1 g/l from 850 to 150 ml/g.

It was noted that when polyelectrolyte dosing was interrupted due to, for example, a pump malfunction, the sludge blanket in the final tank would quickly rise to an unacceptable level.

(vi) Alterations to the flow regime

During the investigation it became apparent that any permanent solution to the bulking problem would require an alteration of the basic treatment process from a 'completely mixed' to a 'plug-flow' type system. Two small-scale pilot plants were commissioned to study the effect of converting the works to a 'plug-flow' system and also to a high-rate oxidation followed by a low rate nitrification two-stage process. Both systems gave encouraging results.

Consideration of the hydraulic situation indicated that the works might most easily be converted in the short term to the two-stage process by simply installing a suitable mobile pump. The conversion is illustrated in Figs 12.1 and 12.2 and operating and loading data before and after conversion are given in Table 12.2.

As a result of this change an immediate improvement in sludge settlement occurred. Typical values of the SSVI before conversion were between 260 and 280 ml/g. The occurrence of no settlement corresponds to an SSVI of 286 ml/g. After modification of the plant the SSVI of the high and low-rate plants dropped to around 100 ml/g. This improvement is shown diagrammatically in Fig. 12.3 where it is also shown that the low-rate plant improved more slowly than the high-rate plant. This was due to the differing sludge ages of the plants which were 2 and 12 days.

The new activated sludges which developed in the two-stage system had fewer filamentous organisms and a larger proportion of flocculating bacteria and ciliated Protozoa than the original sludge.

Large-scale denitrification in the final tanks of the second stage process occurred at times of low BOD loading. To prevent the loss of solids by this route it was decided to reduce temporarily the degree of nitrification achieved in the aeration module by reducing its dissolved oxygen concentration, (2nd to 16th August). At this time a final effluent of almost Royal Commission quality, (20 mg/l BOD, 30 mg/l SS), but high in ammoniacal nitrogen was being produced. It was also observed that when the level of nitrate in the final tanks reached approximately 8 mg/l, denitrification commenced.

One month after the conversion of the plant it was considered that the new activated sludges should be fully established, and that the concentration of dissolved oxygen in the aeration lanes could be increased to allow nitrification to occur. No significant rising sludge problems were encountered, and a good final effluent, low in ammoniacal nitrogen, was produced.

Final effluent quality data are shown in Table 12.3 for periods during which the works were suffering a massive bulking problem (23rd June to 5th July), when measures were taken to control the bulking (6th July to 27th July), and when the bulking was alleviated by the implementation of a two-stage process (2nd August to 3rd September). Average data are given together with 'spot samples' taken by the Thames Conservancy Division. Though the quality of the 24-hour composite samples was poorer than expected during the period when measures were being taken to control bulking, the quality of the effluent discharged to the receiving stream was maintained at Royal Commission standard by the measures described.

Table 12.2 – Operating and Loading Data for Aeration Units (Average figures during the indicated intervals)

Date	Aeration Plant	Feed		MLSS (mg/l)	Sludge Age (d)	BOD Sludge Loading (kg/kg d)	BOD Loading (kg/m³ d)	Flow (m³/d)
		Suspended Solids (mg/l)	BOD (mg/l)					
23rd June to 5th July	Completely mixed	210	346	1455	4.6	0.36	0.52	4106
6th July to 27th July	Completely mixed	235	338	1369	3.9	0.37	0.50	4074
28th July to 29th August	High rate	170	333	1579	2.7	0.72	1.13	3716
	Low rate	79	124	2064	11.5	0.14	0.28	3716

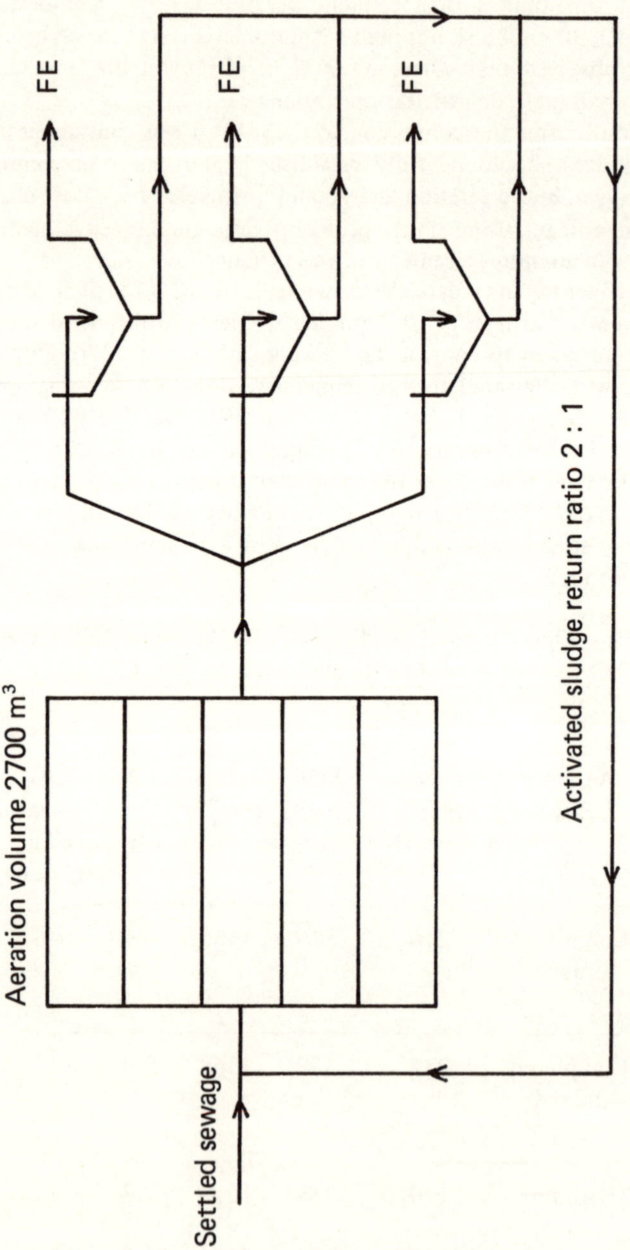

Fig. 12.1 – Flow diagram of Didcot Works before conversion to a two-stage process.

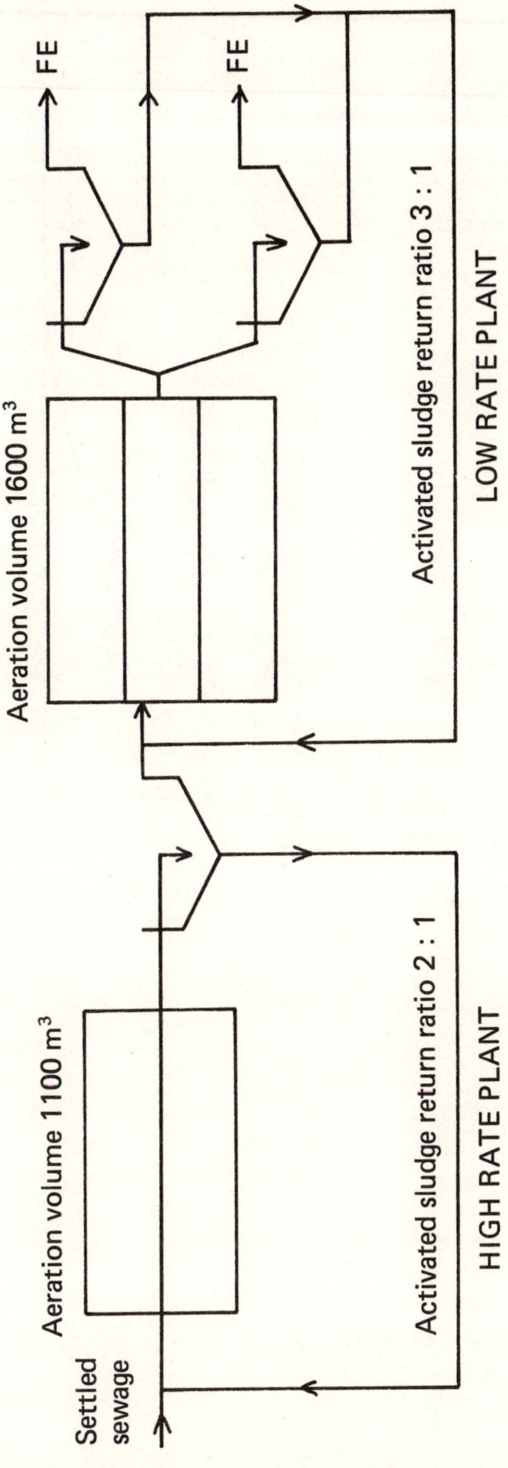

Fig. 12.2 – Flow diagram of Didcot Works after conversion to a two-stage process.

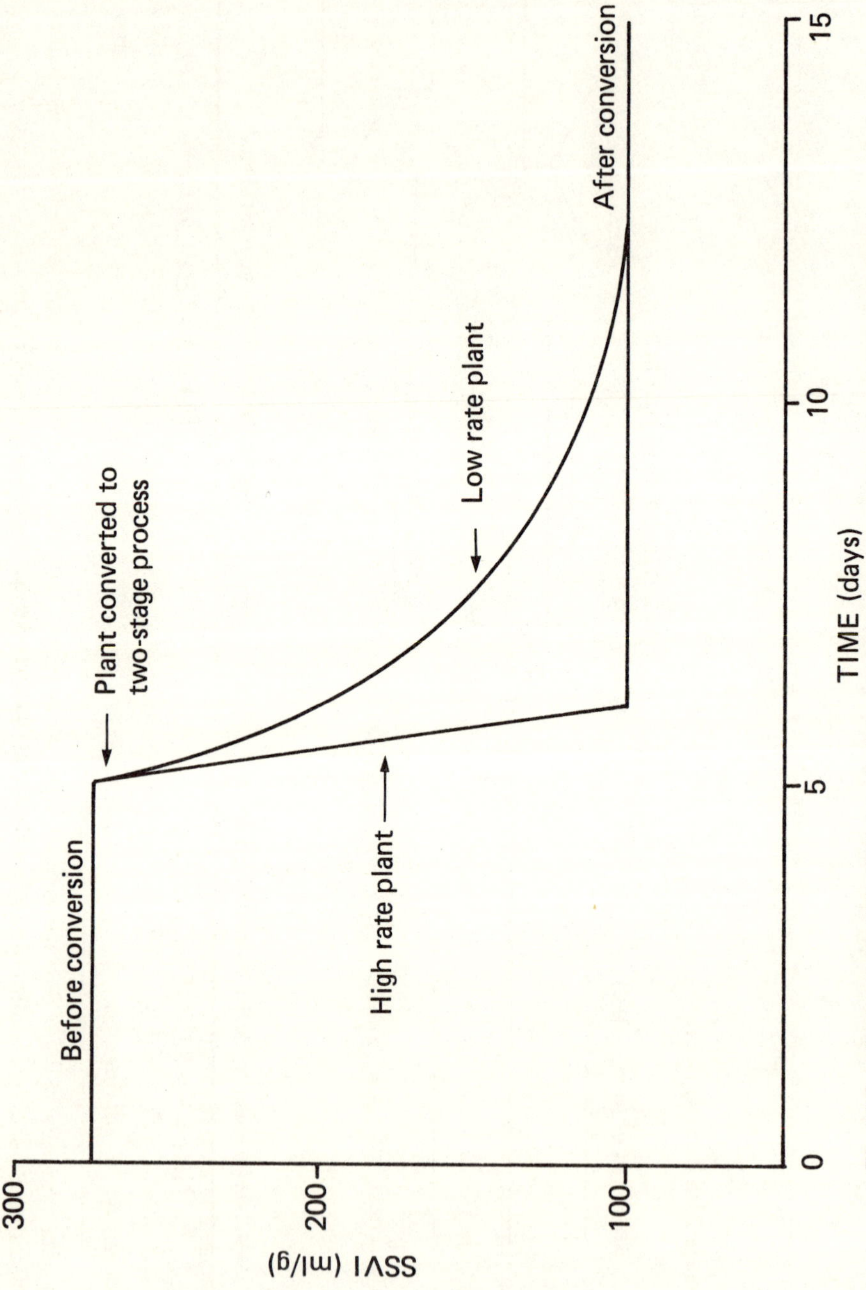

Fig. 12.3 – Sludge settleability before and after conversion of Didcot Works to a two-stage process.

Table 12.3 – Final Effluent (FE) Analytical Data

	SS	BOD	BOD (ATU)	COD	NH_3	NO_3	
Average 23rd June to 5th July	447	253	–	701	4.51	2.55	Massive bulking
Spot samples 2nd July	556	161	164	–	3.1	0.6	
3rd July	812	158	164	–	3.0	0.2	
Average 6th July to 27th July	120	60.9	–	192	4.80	2.71	Control of bulking
Spot samples 7th July	5.2	6.5	2.7	–	11.7	6.1	
24th July	117	75.8		–	1.0	5.9	
Average 28th July to 1st August	109	70		120	3.30	3.30	Two-stage process commissioned but denitrification occurring
Average 2nd August to 3rd September	24	27		74	6.90	6.2	Bulking alleviated by two-stage process
Spot sample 7th August	5.3	9.5			6.4	2.4	

All data expressed in mg/l. Average samples were taken over a 24-hour period by automatic sampler and flow bulked. Spot samples were taken for monitoring purposes by Thames Conservancy Division.

12.4 CONCLUSIONS

- The modification of the sewage works to a two-stage activated-sludge treatment process enabled the plant to treat successfully a sewage containing potato waste, without the use of additional oxygen.
- The most effective method of containing the bulking sludge was found to be dosing the feed to the final tanks with a cationic polyelectrolyte.

It was recommended that the plant should be modified to enable it to run permanently as a 'plug-flow' process. It was considered that this would help not only during the potato season but also throughout the year, as the sludge produced by the 'completely-mixed' system had been demonstrated to exhibit poor settling characteristics.

This modification has since been carried out and the plant has successfully treated the subsequent seasonal loads.

12.5 ACKNOWLEDGEMENTS

The authors thank Mr. J. R. Studholme, Manager, Lambourn Division and his staff for their enthusiastic co-operation in this joint effort. In addition they are grateful for the assistance provided by numerous colleagues, in particular A. J. Rachwal, D. Jesson, Mrs. M. L. Norris and the analytical laboratory staff of Reading Sewage Works and also the staff at WRC Stevenage for helpful discussions.

12.6 REFERENCES

[1] Pike, E. B. and Curds, C. R. The microbial ecology of the activated sludge process, In: *Microbial Aspects of Pollution*, (Sykes, G. and Skinner, F. A. eds), 1971, Society of Applied Bacteriology Symposium, Academic Press, London.

[2] Sezgin, M., Jenkins, D. and Parker, D. S. A unified theory of filamentous activated sludge bulking. *Journal of the Water Pollution Control Federation*, 1978, **50**, 362–381.

Discussion

Chairman: **Professor P. Grau** (Institute of Chemical Technology, Prague, Czechoslovakia)

The discussion was opened by **Mr. E. I. Clark** (North West Water Authority, UK) who commented on the very low ammonia and nitrate–nitrogen concentrations in the final effluent data presented by the authors and asked them for details of the ammonia concentrations in the influent sewage. Mr. Clark also said that the authors had mentioned that it was often difficult to detect positive dissolved-oxygen concentrations in the aeration tank, with the original plant configuration, particularly during periods of peak trade-effluent discharge. He asked the authors if they could provide dissolved-oxygen concentration data for the modified system which allowed comparison with the original configuration. **Mr. Waller** replied that the influent sewage ammonia concentration was approximately 25 mg/l and no problems had been encountered due to nutrient imbalance. He thought that the reason for the low nitrate concentration in the final effluent was the occurrence of denitrification in the aeration tank during periods of very low dissolved-oxygen concentration. In answer to Mr. Clark's second question Mr. Waller replied that the aeration tank dissolved-oxygen profile varied dramatically according to the amount of vegetable canning waste present in the influent sewage. At certain times of the day it was thought that anoxic conditions prevailed throughout most of the aeration tank volume.

Mr. R. A. Varley (Wessex Water Authority, UK) suggested that the authors' investigation of pure oxygen injection as a means of maintaining an adequate dissolved-oxygen concentration had revealed that the amount of oxygen injected was inadequate and that better results might have been obtained if more oxygen had been available. **Mr. Waller** agreed that the amount of oxygen injected was not sufficient to maintain a positive dissolved-oxygen concentration at all times, but this was due to limitations set by the equipment used. However, Mr. Waller went on to say that he considered the problem of poor settleability to be due to the complete-mixing characteristics of the aeration tanks and he was of the opinion that raising the dissolved-oxygen concentration to higher values would not have provided a solution to the bulking problem.

Mr. L. Rennerfelt (Ferriklor International AB, Sweden) said that the experience gained in the treatment of various industrial wastes, such as potato processing wastes, in Scandinavia had shown that sludge settleability problems could, in many cases, be overcome by the addition of ferric salts. Many municipal plants employed chemical dosing continuously and the low dosages required were found to be economically attractive. He asked the authors whether they had considered the addition of ferric salts at the Didcot works. **Mr. Hurley** replied that they had considered such additions but their experiments with lime dosing at high rates (approximately 80 kg/h) had produced no significant improvement in sludge settleability and subsequently they had tended to move away from further consideration of chemical control methods. Calculations of the probable costs involved had reinforced this view. **Mr. Waller** added that chlorine dosing had also been considered but the addition of polymers was found to be much easier to implement.

Mr. E. J. Tomlinson (Water Research Centre) said that the authors had stated that quite early in their investigations it had become apparent that a permanent solution to the problem would require alterations to the plant configuration to achieve a higher degree of 'plug-flow'. He went on to say that the conference proceedings had indicated why such modifications might be considered and he asked the authors why they had decided on such a control strategy at a time when there was much less evidence that it could be successful. **Mr. Waller** replied that the decision had been made after discussions with many of his colleagues in Thames Water Authority and also after consideration of some early work on the subject by the WRC. **Mr. Hurley** said that the bulking problem was so severe that they were quite prepared to perform extensive plant modifications in order to find a permanent solution.

Mr. A. G. Boon (Water Research Centre) said that he shared the authors' belief in the two-stage process and referred to some results he had obtained at the Stevenage Laboratory of the WRC using a pilot-scale two-stage system for the treatment of raw sewage. During an eighteen-month experimental period of operation the sludge in the final stage had always exhibited good settling properties. However, Mr. Boon expressed some reservations about the expense of such a system for the treatment of settled sewage in comparison to conventional processes.

Professor D. Jenkins (University of California, Berkeley, USA) referred to the biological filters which provided secondary treatment at the original Didcot works and asked the authors if they had considered the use of filters, particularly high-rate filters, as the first stage of a two-stage process followed by activated-sludge treatment. **Mr. Hurley** and **Mr. Waller** replied that they had considered this method but decided it was not a feasible proposition at Didcot bearing in mind the urgency of the situation. Mr. Waller added that the existing biological filters at Didcot were on a separate site and the pipeline connecting the two works was sealed soon after commissioning of the new activated-sludge plant.

Mr. L. Rennerfelt said that the use of a high-rate filter before an activated sludge plant was common practice in Sweden for the treatment of strong trade wastes. **Mr. Waller** and **Mr. Hurley** said that in their experience the use of high-rate filtration as a treatment method often gave rise to problems with flies and odour nuisance. They were aware of instances where high-rate filtration had been discounted as a process option because of the odour problem.

Mr. J. L. Hutton (Grampian Regional Council, UK) referred to two works in the Grampian Region which employed high-rate filtration as an initial treatment stage. One of them received settled sewage and a considerable amount of distillery effluent. No odour problems had been reported at this plant. However, the second works treated waste containing a high proportion of slaughterhouse effluent and this had given rise to severe odour problems. Mr. Hutton was of the opinion that the severity of the problem depended on the type of waste being treated.

Mr. Hutton also mentioned the sludge settleability problems encountered at Persley STW. Extensive pilot-scale investigations had been performed and none of the control methods employed, including the use of a 'plug-flow' configuration, could provide a solution, except the 'Unox' pure oxygen system. A full-scale 'Unox' plant had been installed on the site and no problems with bulking had arisen since that time. Mr. Hutton added that investigations using the 'Vitox' system of oxygen injection had not cured the problem of poor settleability. **Mr. Hurley** said that Thames Water Authority had had a similar experience with the treatment of a brewery waste and had reached the same conclusion. A full scale 'Unox' plant had recently been built to treat the waste.

Mr. J. O'Neill (Yorkshire Water Authority, UK) said that most activated-sludge plants suffered settleability problems at some time. He related an instance where the loss of solids from a plant had been prevented during a period of temporary poor settleability by diverting the effluent to storm water tanks. He had calculated that the storm water tanks were only in use as such for about 5 per cent of the year and he thought that it would be desirable for plant designers to consider the use of storm water tanks as tertiary settlement tanks. **Mr. Hurley** said that they had used the storm water tanks at Didcot to try to prevent solids loss but the bulking problem was so severe that they were quickly filled to capacity.

Chapter 13

The application of settleability tests for the control of activated-sludge plants

A. J. Rachwal, D. W. M. Johnstone, M. J. Hanbury, and **D. J. Critchard,** Thames Water Authority

13.1 INTRODUCTION

A long-term study of several Carrousel activated-sludge plants has been completed recently by Thames Water Authority. Problems with bulking were encountered early in the work necessitating a detailed investigation of sludge settleability and factors that could influence it. The traditional technique of measuring sludge settleability, Sludge Volume Index (SVI), was compared with the more recent Stirred Specific Volume Index (SSVI) [1, 2]. Full-scale experiments, monitoring a final settlement tank under various solids loadings, were carried out to assess Mass Flux Theory and SSVI as practical techniques for the operation and design of final settlement tanks. This chapter describes some of the more significant results obtained between 1976 and 1980. Comprehensive details of the work have been published in earlier papers [3, 4].

13.2 EXPERIMENTAL

13.2.1 Laboratory procedure

Samples of mixed liquor and returned sludge were taken daily and subjected to laboratory examination within 30 minutes of sampling. Delay in examination was found to give poor reproducibility, particularly with regard to settlement characteristics. Sludge settling characteristics were determined using both SVI and SSVI techniques. Hindered settling velocities for use in Mass Flux Theory were determined concurrently with SSVI measurements in the stirred cylinders. All other analyses were carried out using the standard methods used in the UK Water Industry. A total of some 2000 samples taken over a four-year period at Ash Vale Works, and some 700 samples over a two and a half-year period at Cirencester Works provided the data for the comparison of SVI and SSVI techniques.

13.2.2 Final settlement tank monitoring

Sludge/water interface depths were determined using a light-cell sludge level detector at a minimum of six positions across the scraper bridge. The rotating bridge was stopped for five minutes during the measurements to avoid 'wave' effects at the sludge/water interface. During the storm simulations measurements were made every 15 to 30 minutes. Samples of sludge were taken at various depths using a simple corked bottle which was opened at the required depth. The performance of the final settlement tank was studied during a series of eleven storm periods. Four were naturally-occurring storms and seven were artificially simulated at controlled flow rates. To achieve the required high flows, up to four times Dry Weather Flow (DWF), final effluent was recycled into the inlet sewer. Figure 13.1 shows the Ash Vale Carrousel site layout.

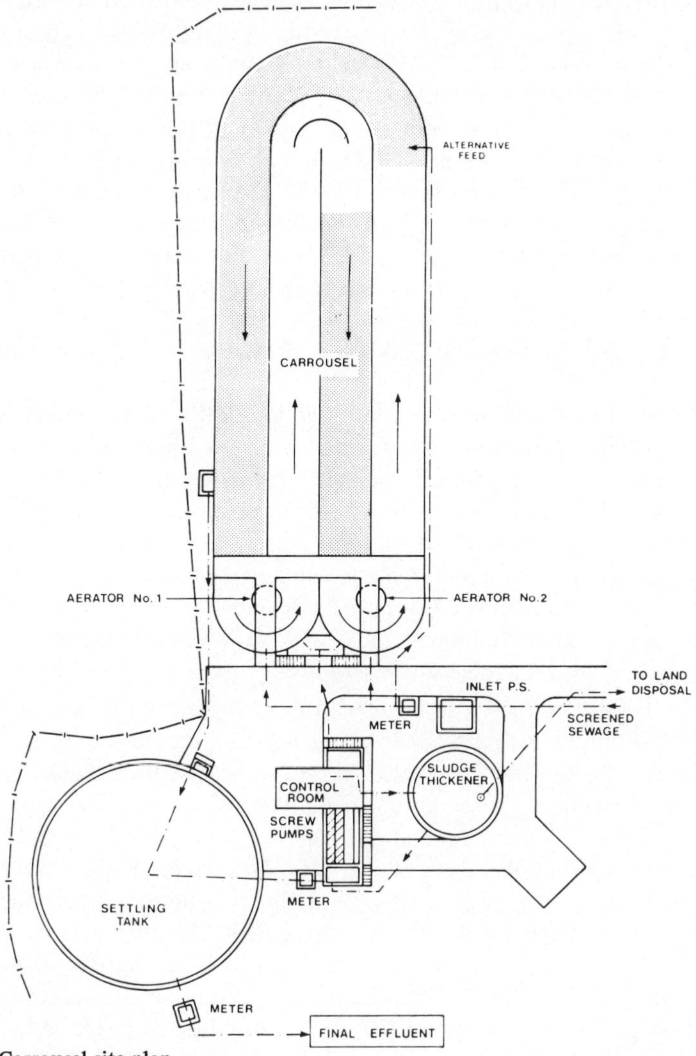

Fig. 13.1 – Carrousel site plan.

13.3 RESULTS AND DISCUSSION

13.3.1 Effect of feed arrangements on SSVI

SSVI is now used at many activated-sludge plants within Thames Water Authority as the principal operational parameter for measuring sludge settleability. Experience has shown that the objectives set by the Water Research Centre in designing the apparatus have been achieved. By way of illustration, two examples of operational experiences are presented in Figs 13.2 and 13.3. Figure 13.2 refers to Ash Vale Works and Fig. 13.3 to Cirencester Works.

In practice, SSVI provides a measurement of settleability with a range from about 40 ml/g to 286 ml/g. A sludge with an SSVI of less than 40 ml/g would have to be capable of being thickened to a concentration greater than 2.5 per cent in 30 minutes, which is unlikely even for the most readily settleable sludges. Bulking sludges have SSVI values greater than 120 ml/g, and a sludge which did not settle at all in 30 minutes would have an SSVI, at 3.5 g/l, of 286 ml/g.

Early in 1977 problems with bulking sludge were encountered at the Ash Vale Carrousel. SSVIs gradually rose, from less than 100 up to 140 ml/g, and some washout of the final tank sludge blanket occurred during periods of high flow. Various strategies were discussed in an attempt to find a solution to the problem.

Mass Flux Theory and SSVI can predict the maximum solids handling capacity (expressed as kg solids/m^2h) of a final settlement tank, a detailed discussion of which is given in a later section of this chapter. Reducing the flow to a final tank or lowering the Mixed Liquor Suspended Solids (MLSS) concentration can reduce the applied loading to below the theoretical maximum. Both techniques were used at Ash Vale with some measure of success, flow being reduced by 15 per cent and MLSS by 40 per cent. However, most sewage works cannot divert flow to another plant and too severe a reduction in MLSS can result in loss of nitrification due to the lowering of sludge age and washout of nitrifying organisms. No improvement in SSVI was observed over a period of five sludge ages after the rapid reduction in MLSS.

Calculation of the critical underflow rate indicated that no benefit could be obtained from increasing the rate of return from the final tank as it was already operating close to the critical value for such a poor settling sludge. A cure, rather than a suppression of the symptoms, was required for the bulking problem.

Bearing in mind the finding of WRC [5], Chudoba [6] and Thames Water Authority [7], that plants employing 'plug-flow' configurations were less susceptible to bulking than plants which employ 'completely-mixed' configurations, methods were sought to increase the 'plug-flow' characteristics of the Carrousel system. The aeration tank can be considered to be 'completely-mixed' but with some 'plug-flow' attributes. The circulation period is of the order of 15 minutes, based on distinct peaks observed from point additions during radiotracer trials, and the hydraulic residence time is about 30 hours, based on sewage flow. Hence the tank contents circulate 120 times during one residence time. Viewed over time scales in excess of 10 per cent of the residence time, the plant can be considered to be 'completely-mixed', whilst viewed over short periods of less than one hour, it can be considered to be 'plug-flow'. The system is further complicated by the fact that during one 15-minute circulation period a plug of mixed liquor passes through two aerobic and two anoxic zones as illustrated in Fig. 13.1.

Until April 1977, both crude sewage and return activated sludge were fed equally into the two aeration zones. It was considered that introduction of the total feed into one aeration zone only would increase the tank's 'plug-flow' characteristics. This change was carried out on the 5th April 1977. It can be seen from Fig. 13.2 that no change was observed in the SSVI for a period of one sludge age (13 days). However, a rapid fall in SSVI from 140 ml/g to 90 ml/g then occurred over a further period of one sludge age. During the period following the change in feed configuration, the MLSS concentration was maintained relatively constant, and there was no significant change in loading, (0.06 kg BOD/kg MLSS d), or sludge age.

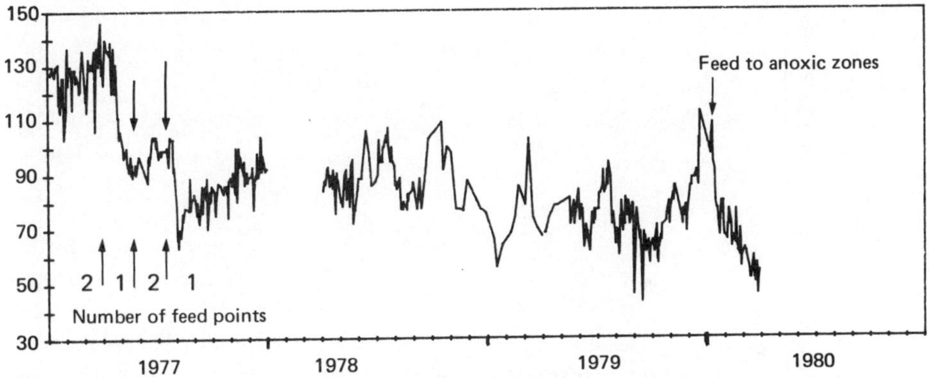

Fig. 13.2 – SSVI for Ash Vale Carrousel 1977–80.

Having operated with a single feed for three sludge ages the original feed pattern was re-adopted and SSVI very gradually increased from 90 mg/l to 105 ml/g. After operating with the original two-position feed for four sludge ages, the single feed arrangement was again adopted (July 1977). As before, no change was observed for one sludge age. Then a rapid fall in SSVI, from 105 ml/g to 65 ml/g, occurred during a further sludge age. The MLSS was not as constant as for the first change in feed arrangements, being reduced by 20 per cent over a period of one week just before the feed change was made. Hence sludge age was somewhat reduced and loading increased at the time of this last change. The single feed arrangement into an aeration zone was maintained for over two years following these experiments. SSVIs remained in the range 60 to 100 ml/g with no further problems of sludge blanket loss from the final tank.

In December 1979, following experimental changes which had halved the aeration tank retention time and doubled the sludge loading, SSVIs began to rise above 100 ml/g and sludge blanket spills started to occur at high flow. The opportunity to test the effect of feeding all the sewage and return sludge into an anoxic rather than aerobic zone was taken. This was done in January 1980 and it can again be seen in Fig. 13.2 that the SSVI fell dramatically after one sludge age (now seven days), from 110 ml/g to 50 ml/g. The single feed into an aerobic zone was readopted during February but no change in SSVI was observed. At the current BOD sludge loading of 0.08 to 0.10 kg/kg d, the plant is operating at an SSVI of 50 to 70 ml/g, which is a very good settling sludge.

These observations provide further evidence that altering feed arrangements to influence the 'plug-flow' nature of an aeration tank and/or the presence of anoxic zones can affect sludge settlement characteristics significantly. It is suggested that a sudden change in these aeration tank feed arrangements can improve sludge settlement characteristics by reducing the number of bulking organisms. Reverting to the original feed conditions may be possible once the bulking organisms are removed from the system.

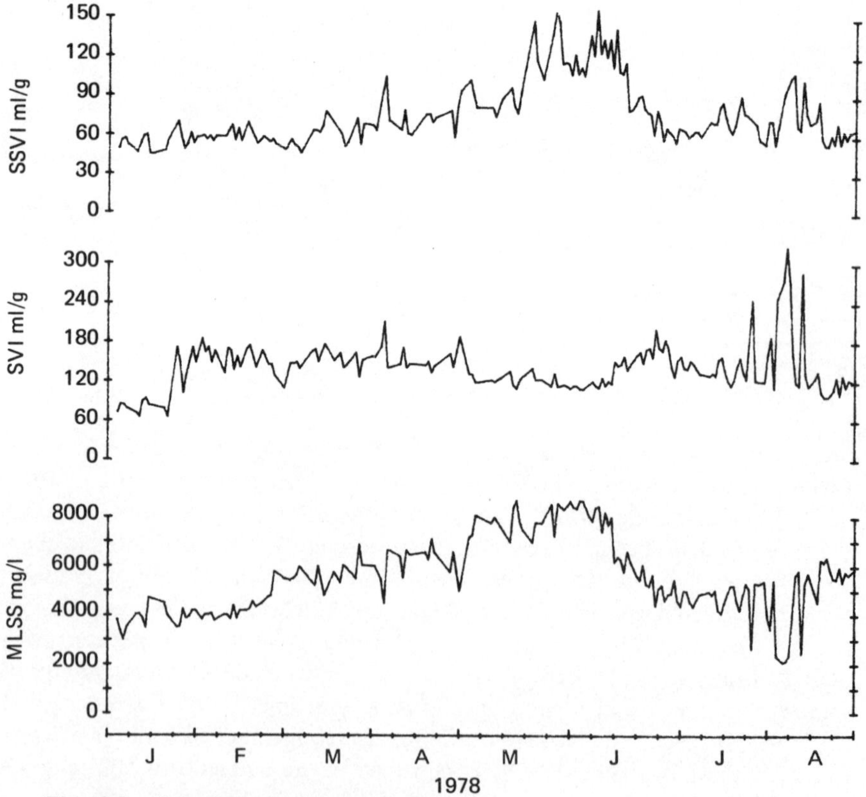

Fig. 13.3 Data from the Cirencester Carrousel.

13.3.2 Comparison of SVI and SSVI

During the periods of changing sludge settleability the results indicated that SSVI was more effective in measuring these changes than was SVI, principally because of the independence of SSVI on solids concentration. Data from the Cirencester Carrousel (Fig. 13.3) also show how a marked change in settleability was readily reflected by SSVI but not at all by SVI. From the end of February until early June 1978, there was a large increase in MLSS arising from an operational problem with sludge disposal. By the beginning of May the mass of sludge held in the aeration tank was so large that the installed aeration equipment was not sufficient to meet the oxygen demand and sludge settleability deteriorated. Examination of Fig. 13.3 shows how this deterioration was readily reflected by the SSVI but not by the SVI. The reason was that the 30 minute settled volume in the unstirred, litre cylinder increased at approximately the same rate as the increase in MLSS.

The above examples highlight one of the main deficiences in SVI; its dependence on suspended solids. Another way of illustrating this is given in Fig. 13.4.

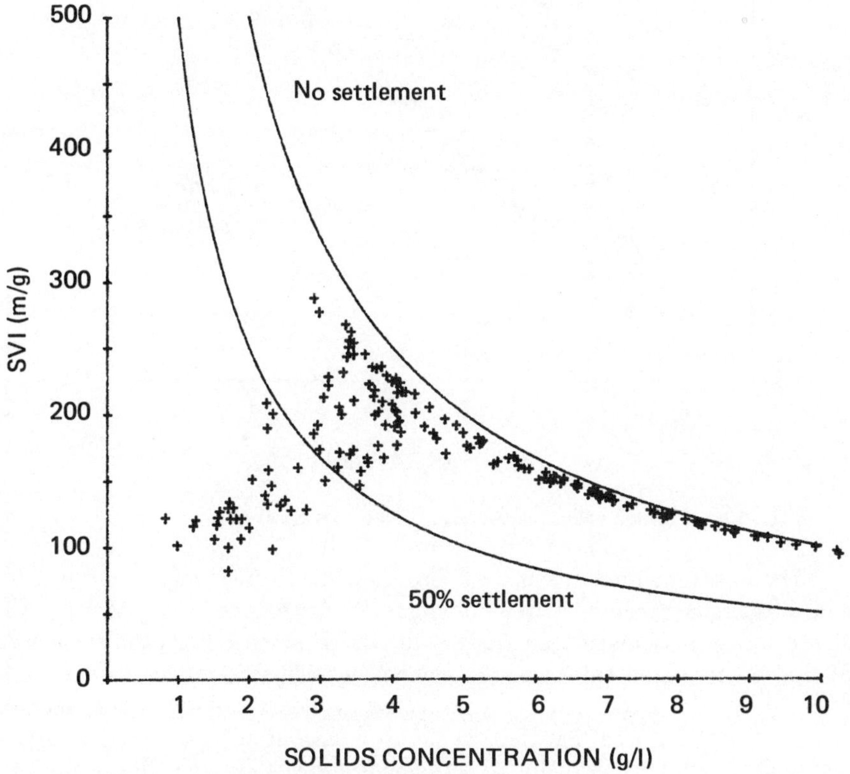

Fig. 13.4 – SVI v. solids concentration (80 < SSVI < 85).

To construct the figure, all the data within a narrow band of SSVI were selected and the corresponding SVIs plotted against solids concentration. Within such a narrow band of SSVI all the sludges had essentially the same settleability, yet the SVIs covered a range from about 80 to 290 ml/g. Also up to a solids concentration of 4 g/l, there was a steady rise in SVI. At concentrations above 4 g/l there was virtually no settlement in the unstirred cylinder which means that the maximum useful range of SVI is to a concentration slightly less than 4 g/l.

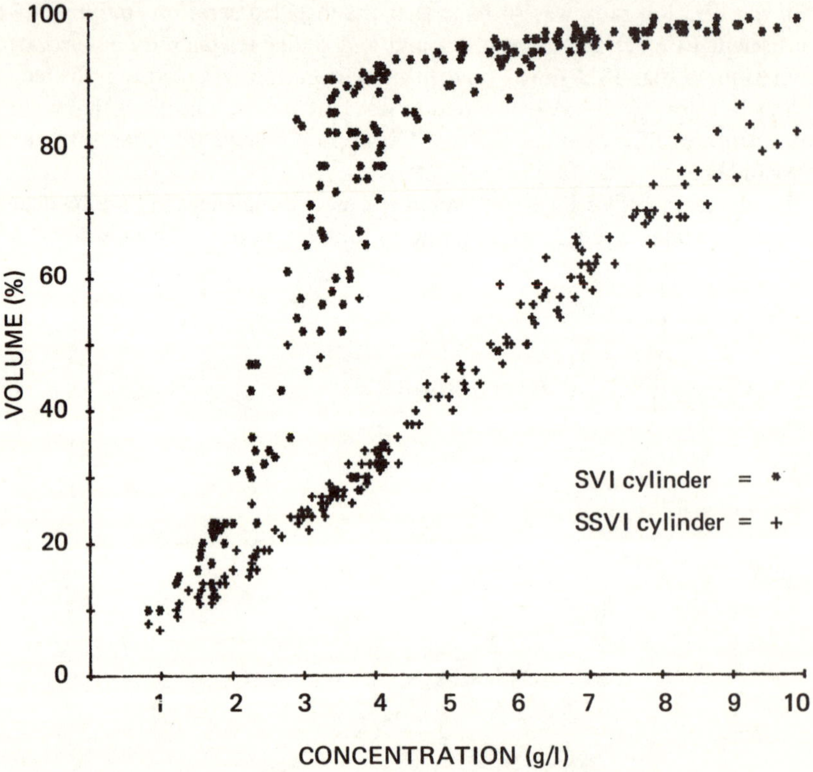

Fig. 13.5 – Per cent volume v. concentration (80 < SSVI < 85).

The data presented in Fig. 13.5 also show the limited working range of SVI and, at the same time, show the much larger working range of SSVI. As in Fig. 13.4 all the data were obtained from sludges with the same settleability (SSVIs between 80 and 85 ml/g). In this case, the percentage volume occupied by the sludge after 30 minutes settlement in the respective cylinders is plotted against concentration. It can be seen that there is almost a discontinuity in the SVI curve at just under 4 g/l, with virtually no settlement above 4 g/l. By contrast, the SSVI curve is only slightly sigmoidal over the whole concentration range. A series of these curves shows that the rate of increase of per cent volume in the SSVI cylinder tends to decrease at higher solids concentrations as the settleability deteriorates. Even with bulking sludges (SSVI 125 to 130 ml/g) the parameter is effective up to a concentration of 7.5 g/l.

It is important to note that, over the range of solids concentrations normally encountered in activated-sludge plants, the relationship between settled volume in the SSVI cylinder and concentration is very close to linear. This justifies the technique of interpolating the measured values to provide a standard value at 3.5 g/l.

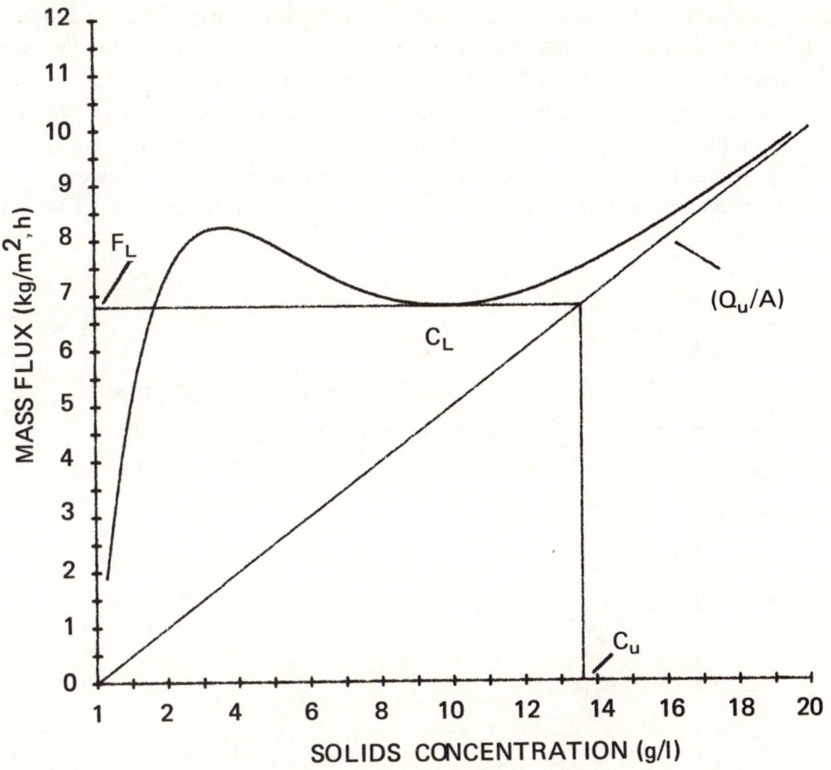

Fig. 13.6 – Total mass flux v. solids concentration.

13.3.3 Mass flux analysis and SSVI

Mass flux analysis examines the flow of solids passing through each cross-section of the settlement tank and considers that the total flux of solids per unit surface area is comprised of two components; one due to gravitational settlement of the sludge, and one due to withdrawal of sludge by continuous underflow from the base of the tank. Thus the total mass flux (F) is given by:

$$F = VC + \frac{Q_u}{A} C \tag{13.1}$$

Where V is the hindered settling velocity (m/h)

C is the initial concentration of solids (kg/m^3)

Q_u is the underflow rate (m^3/h)

A is the cross-sectional area of the tank (m^2),

As long as the underflow rate is not too high, the curve of F against C will have a maximum followed by a minimum as shown in Fig. 13.6. The value of F at the minimum, F_L, determines the maximum predicted solids loading which can be applied to the tank before solids are lost in the overflow.

The results presented in this chapter arise from the analysis of 733 hindered settling velocities measured during the investigations. A plot of the hindered settling velocities against concentration is given in Fig. 13.7. However, these results cover a range of sludge settleabilities (SSVI 57 to 146 ml/g) which is

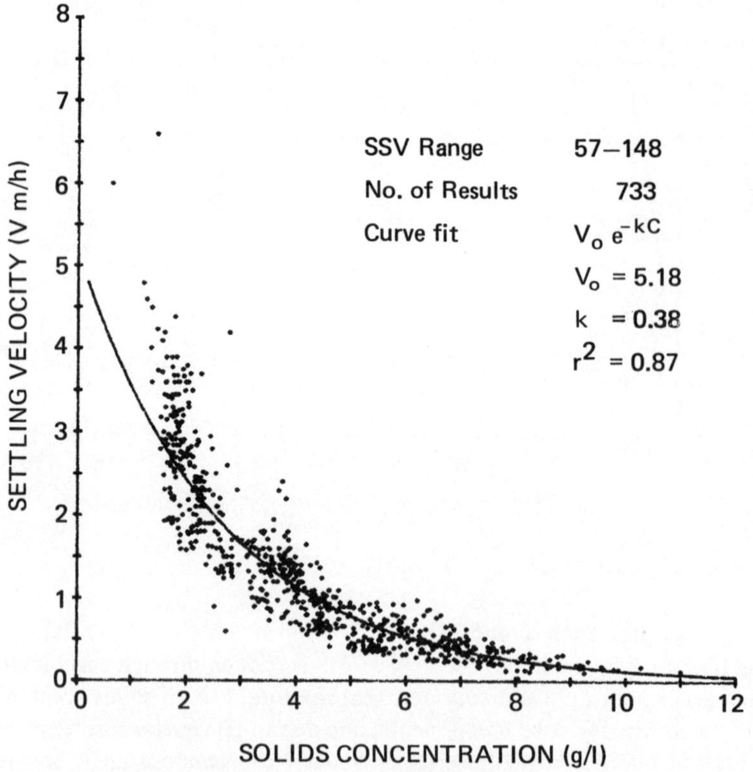

Fig. 13.7 – Hindered settling velocity v. solids concentration.

considered to be too large to provide an accurate analysis of a final tank operating with a sludge of given settleability. To obtain more precise information the data were split into four SSVI ranges, 60 to 79, 80 to 99, 100 to 119 and 120 to 139 ml/g, and the analysis performed for each range.

To facilitate an analytical approach to solving Equation 13.1 for a given value of $\frac{Q_u}{A}$, curves of hindered settling velocity against concentration were expressed in the exponential form adopted by White [1,2].

$$V = V_0 e^{-kC} \tag{13.2}$$

so that Equation 13.1 could be written as

$$F = V_0 C e^{-kC} + \frac{Q_u}{A} C \tag{13.3}$$

where V_0 and k are arbitrary constants. V_0 can be considered to be the settling velocity of an individual floc particle at infinite dilution. Regression analysis, using the exponential expression, provided values for the arbitrary constants. Regression analyses were also carried out using power law, logarithmic and hyperbolic expressions, but from comparison of the coefficients of correlation (r^2) their fits were not as good as that of the exponential function.

Table 13.1 – Curve fits of hindered settling velocity against concentration

SSVI range (ml/g)	Number of results	Solids concentration range (g/l)		Curve fits on data					
				Exponential $V = V_0 e^{-kC}$			Power	Log	Hyp
	n	Min.	Max.	V_0	k	r^2	r^2	r^2	r^2
60 - 79	83	0.65	9.80	6.500	0.3588	0.9607	0.9165	0.9095	0.8611
80 - 99	310	1.23	11.0	5.826	0.3712	0.9591	0.9367	0.8557	0.8991
100 - 119	148	1.55	9.62	5.300	0.4067	0.9374	0.9002	0.9254	0.9337
120 - 139	168	1.53	8.31	4.234	0.4222	0.9618	0.9417	0.9211	0.9392
57 - 146	733	0.65	11.0	5.182	0.3847	0.8732	0.8673	0.8013	0.8369

The results of the curve fitting analyses are given in Table 13.1 which also gives r^2 values for the three other expressions. The values of V_0 and k were used in Equation 13.3 to compute solids flux analyses for each SSVI range, at underflow velocities from 0.13 to 0.75 m/h (0.5 to 3 DWF) and the results are presented in Table 13.2. Two graphical examples representing analyses at underflow velocities of 0.25 and 0.75 m/h (1 and 3 DWF) are given in Figs 13.8 and 13.9

respectively. A comparison of the graphs shows how solids handling capacity increases with increased underflow rate, and it can be seen (Fig. 13.9) that, for the higher ranges of SSVI, there are no minima in the curves, only points of inflexion. Hence, if the tank was to be operated with an underflow rate of 0.75 m/h, it would no longer behave as a thickener for sludges with SSVIs greater than about 100 ml/g. It would merely act as a clarifier, because the underflow velocity is above the critical value ($Vo\ e^{-2}$) as given by White [2]. In this case the maximum solids loading is given by solving Equation 13.1 for the inlet concentration.

$$F = V_f C_f + \frac{Q_u}{A} C_f \qquad (13.4)$$

Where V_f is the settling velocity of the MLSS with concentration C_f.

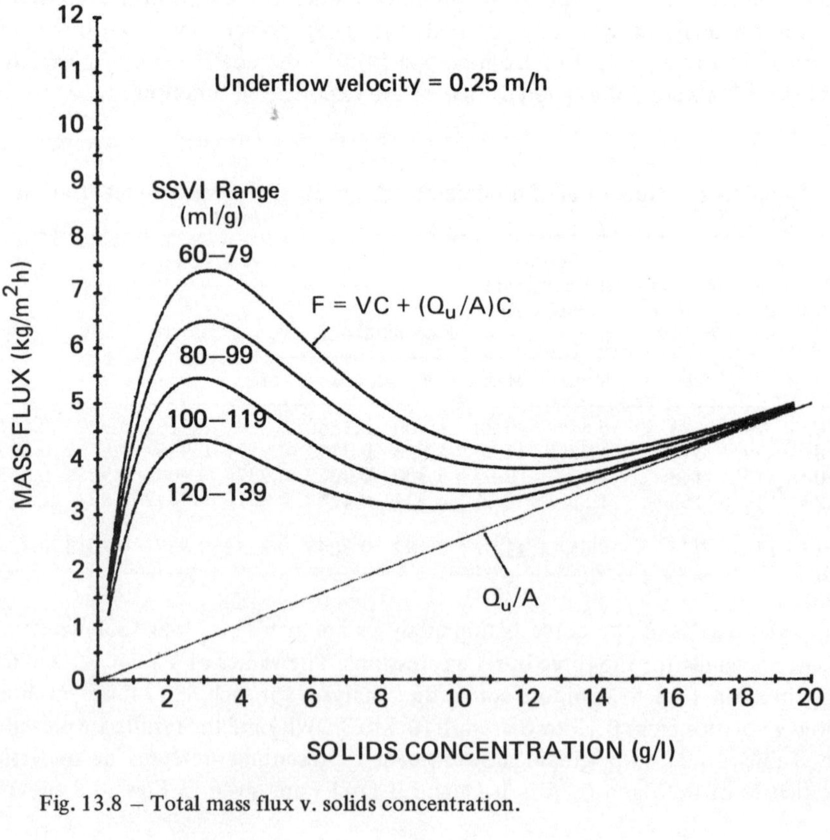

Fig. 13.8 – Total mass flux v. solids concentration.

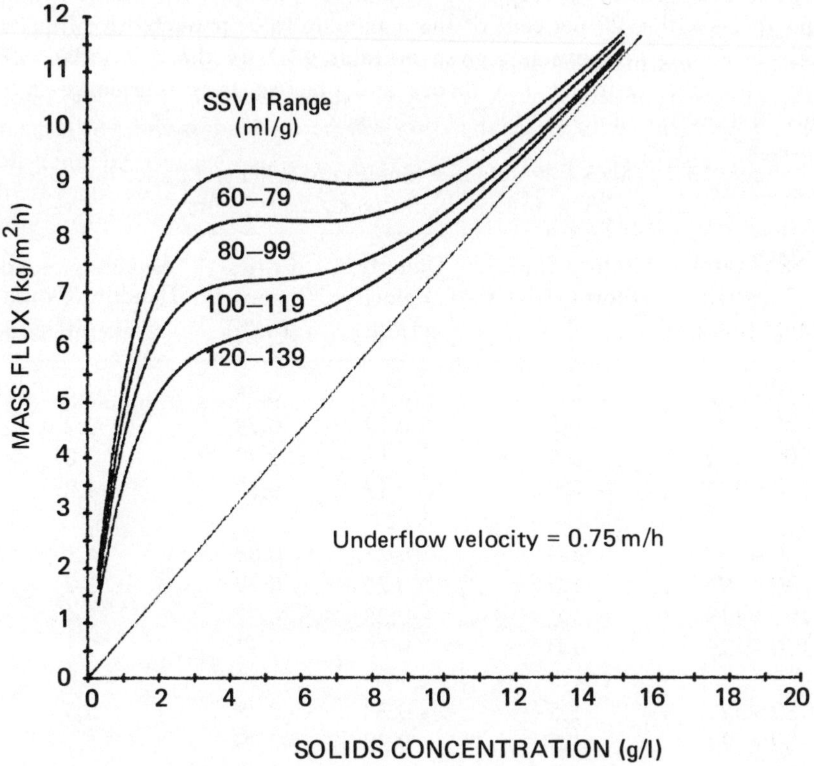

Fig. 13.9 – Total mass flux v. solids concentration.

13.3.4 Mass flux and SSVI in the evaluation of settlement tank performance

During four naturally-occurring storms, sludge was washed-out into the final launder. In these storms, flows between 3 and 4 DWF were experienced for up to nine hours. To study sludge wash-out in detail, high storm flows were simulated by returning final effluent to the works inlet. Three studies were carried out with a double-weir configuration on the effluent launder, and four with the outer weir blanked off. It was found that the original design, with a double-sided launder, accelerated sludge wash-out when the sludge layer was near the base of the launder. Sludge was only washed-out over the outer layer and details of this have been reported elsewhere [3]. When the sludge layer was more than 20 cm below the base of the launder the rates of sludge/water interface rise were essentially the same for both configurations. The data pertaining to all the storm conditions are given in Table 13.3.

In all cases when the sludge was washed out, the applied solids loading was found to be within 20 per cent of the maximum value predicted for Equation 13.3, when using the constants given in Table 13.2 for the appropriate SSVI range. This agreement between theory and practice gives confidence in the ability of mass flux to predict tank performance.

Table 13.2 – Mass flux data as a function of SSVI and underflow rate
DWF = 4545m^3/d Surface Area = 755m^2

SSVI (at 3.5 g/l) Range (ml/g)	Return Sludge Flow (xDWF)	Underflow Velocity (m/h)	Critical Underflow (m/h)	Maximum Solids Handling Capacity (kg/m^{2}h)
60 - 79	0.5	0.13	0.88	2.4
80 - 99	0.5	0.13	0.79	2.2
100 - 119	0.5	0.13	0.72	2.0
120 - 139	0.5	0.13	0.57	1.8
60 - 79	1.0	0.25	0.88	4.1
80 - 99	1.0	0.25	0.79	3.8
100 - 119	1.0	0.25	0.72	3.4
120 - 139	1.0	0.25	0.57	3.1
60 - 79	1.5	0.38	0.88	5.5
80 - 99	1.5	0.38	0.79	5.1
100 - 119	1.5	0.38	0.72	4.6
120 - 139	1.5	0.38	0.57	4.1
60 - 79	2.0	0.50	0.88	6.8
80 - 99	2.0	0.50	0.79	6.4
100 - 119	2.0	0.50	0.72	5.7
120 - 139	2.0	0.50	0.57	5.0
60 - 79	2.5	0.63	0.88	8.0
80 - 99	2.5	0.63	0.79	7.4
100 - 119	2.5	0.63	0.72	6.5
120 - 139	2.5	0.63	0.52	*
60 - 79	3.0	0.75	0.88	9.0
80 - 99	3.0	0.75	0.79	8.3
100 - 119	3.0	0.75	0.72	7.2
120 - 139	3.0	0.75	0.52	*

* Exceeds critical underflow velocity.

Table 13.3 – Applied and predicted solids loading on the FST when sludge layer washover was occurring

Date	Mixed Liquor			Flow (x DWF) to the Final Settling Tank		Flow/Area (m/h) in the FST		Solids Loading on FST $kg/m^2 h$		Notes. unless stated otherwise double sided weir
	SS g/l	SSVI (at 3.5 g/l) (ml/g)	Hindered Settling Velocity (m/h)	Crude Sewage	Return sludge	Total $\frac{Q_o}{A} + \frac{Q_u}{A}$	Underflow $\left(\frac{Q_u}{A}\right)$	Applied	Max Predicted Table 13.2	
29.11.76	4.15	100	1.0	3.1 – 4.0	2.0	1.3 – 1.5	0.50	5.4 – 6.2	5.7	Natural storm
30.11.76	4.05	106	1.0	3.0 – 4.3	2.0	1.2 – 1.6	0.50	5.6 – 6.4	5.7	Natural storm
21.12.76	5.20	114	0.5	1.5 – 1.7	1.8	0.8 – 0.9	0.45	4.3 – 4.6	5.7	Natural storm
13. 1.77	5.40	130	0.4	1.9 – 2.7	1.8	0.9 – 1.1	0.45	5.0 – 6.0	5.0	Natural storm
7.12.76	4.35	101	0.9	2.0 – 2.4	2.1	1.0 – 1.1	0.52	4.4 – 4.9	5.7	Simulated storm
9.12.76	4.40	106	0.9	2.6 – 3.2	2.1	1.2 – 1.3	0.52	4.8 – 6.3	5.7	Simulated storm
1. 4.77	4.10	123	0.9	2.6	2.0	1.15	0.50	4.7	5.0	Simulated storm
2. 5.77	3.35	101	1.5	2.9	1.4	1.1	0.35	3.7	4.6	Simulated storm 1 weir only
3. 5.77	3.50	100	1.3	3.6	2.9	1.6	0.73	5.6	7.1	Simulated storm 1 weir only
13. 4.77	3.85	139	0.8	2.6	2.6	1.30	0.65	5.0	5.6*	Simulated storm 1 weir only
4. 5.77	3.40	96	1.4	2.9	2.9	1.45	0.73+	4.9	5.4+	Simulated storm 1 weir only

* Above critical underflow Equation 13.4 used.
\+ At critical underflow.

Table 13.4 – Maximum MLSS Levels to Prevent Sludge Washout

SSVI (at 3.5 g/l) Range (ml/g)	Return Sludge Flow		Maximum Predicted Solids Handling Capacity	Maximum Maintainable MLSS Levels (g/l) for Maximum Flow of:–				
				2 DWF	2.5 DWF	3 DWF	3.5 DWF	4 DWF
	xDWF	$\frac{Q_u}{A}$	(kg/m²h)	$\frac{Q_o}{A} = 0.5$	$\frac{Q_o}{A} = 0.625$	$\frac{Q_o}{A} = 0.75$	$\frac{Q_o}{A} = 0.875$	$\frac{Q_o}{A} = 1.0$
60 - 79	0.5	.125	2.4	3.75	3.00	2.75	2.40	2.10
80 - 99	0.5	.125	2.2	3.50	2.75	2.50	2.20	1.90
100 - 119	0.5	.125	2.0	3.00	2.50	2.25	2.00	1.75
120 - 139	0.5	.125	1.8	2.75	2.25	2.00	1.80	1.60
60 - 79	1.0	.25	4.1	5.50	4.75	4.25	3.75	3.25
80 - 99	1.0	.25	3.8	5.00	4.50	4.00	3.50	3.00
100 - 119	1.0	.25	3.4	4.50	4.00	3.50	3.00	2.50
120 - 139	1.0	.25	3.1	4.00	3.50	3.00	2.50	2.00
60 - 79	1.5	.375	5.5	6.00	5.50	4.75	4.25	4.00
80 - 99	1.5	.375	5.1	5.50	5.00	4.50	4.00	3.50
100 - 119	1.5	.375	4.6	5.00	4.50	4.00	3.50	3.25
120 - 139	1.5	.375	4.1	4.50	4.00	3.50	3.25	3.00
60 - 79	2.0	.5	6.8	7.00	6.00	5.50	5.00	4.50
80 - 99	2.0	.5	6.4	6.50	5.50	5.00	4.50	4.00
100 - 119	2.0	.5	5.7	5.50	5.00	4.50	4.00	3.50
120 - 139	2.0	.5	5.0	5.00	4.50	4.00	3.50	3.00
60 - 79	2.5	.625	8.0	7.00	6.50	5.75	5.25	4.75
80 - 99	2.5	.625	7.4	6.50	6.00	5.25	4.75	4.25
100 - 119	2.5	.625	6.5	5.75	5.00	4.50	4.25	3.75
60 - 79	3.0	.75	9.0	7.00	6.50	6.00	5.50	5.00
80 - 99	3.0	.75	8.3	6.50	6.00	5.50	5.00	4.50
100 - 119	3.0	.75						
120 - 139	3.0	.75						

Having demonstrated agreement between theory and practice, calculations using mass flux (Equation 13.3) can be made over a range of SSVIs to provide the plant operator with a set of constraints or guidelines outside of which the plant should not be operated. Such a set of constraints is shown in Table 13.4 This gives the maximum MLSS (C_f) values which can be maintained in the aeration tank to prevent the applied solids loading $\left(\frac{Q_o}{A} + \frac{Q_u}{A}\right) C_f$ from exceeding the solids handling capacity of the tank. The data encompass the whole range of SSVI and return sludge flow rates likely to be encountered at the plant under examination. Although the guidelines in Table 13.4 are specific to the settlement tank in question, it is a relatively easy task to construct a similar table for other tanks by making measurements of hindered settling velocities within known SSVI ranges.

The results of these studies are in general agreement with those of White [1,2] and further support the claims that mass flux theory can be used as a rational basis for the design of settlement tanks for activated sludge [8,9]. In fact, a number of settlement tanks, currently being constructed by Thames Water Authority, have been designed using this approach.

While mass flux theory provides a rational basis for the selection of surface area and underflow rates in its present form, it has other limitations. Mass flux does not always predict the concentration of solids in the underflow. In the present studies, the concentration of solids in the underflow was always much less (30 to 40 per cent) than that predicted by theory; similar phenomena were found by Hibberd and Jones [10]. It is believed that, in this work, this is due to the shallowness of the tank preventing the solids from settling to a zone of compression. Considerable concentration gradients were found in the sludge layer, indicating that the sludge was in a hindered settling zone with minimal compression forces. This suggests that tank depth is important in establishing solids concentration in the underflow. The effect of depth on thickening is a feature of tank performance not considered by mass flux theory, although one attempt has been made to account for it [10].

It is also important to provide sufficient depth to accommodate fluctuations in the level of the sludge/water interface caused by variation in flow and solids loading. This is another aspect of tank performance not considered by mass flux theory, or most other design philosophies.

13.3.5 Sludge blanket level control in the settlement tank

Figure 13.10 illustrates the variation of sludge blanket level with flow over a 48-hour period in the Ash Vale settlement tank. Equipment had been installed to reduce this variation by linking the rate of return to the rate of incoming sewage flow. The control system consisted of a motorized gate valve in the settlement tank underflow outlet pipe, feeding the base of a return sludge screw pump. A linearizing logic circuit linked the underflow valve setting to the incoming sewage flow. Unfortunately the control valve was both fast acting and sensitive to blockages when partially throttled, tending to hunt from fully open to fully closed. This control system was abandoned at Ash Vale with the recommendation that either a slower acting motorized valve or a motorized bellmouth be installed.

The latter type of control valve was utilized for an automatic sludge blanket level control system installed for Thames Water's latest Carrousel at Leatherhead. The plant was completed in September 1980 and incorporated variable output, pre-rotation return sludge pumps that should give energy savings when return rates are low. The return rate was determined by motorized, sludge blanket, photocell detectors installed on the scraper bridge of each final settlement tank. This type of control sensor was preferred to one linking return rate to incoming flow as it could also provide automatic warning of an imminent blanket spill.

At the time of this chapter's preparation (November 1980) this blanket level control system had yet to be commissioned and it was expected that the weakest link would be the photocell blanket detector. A study is currently being made of the reliability of various sludge blanket detectors to determine operating procedures with regard to factors such as frequency of sensor cleaning and sensitivity. The authors are also investigating a method which could be used to determine the rate of rise of the sludge/water interface during periods of solids overload.

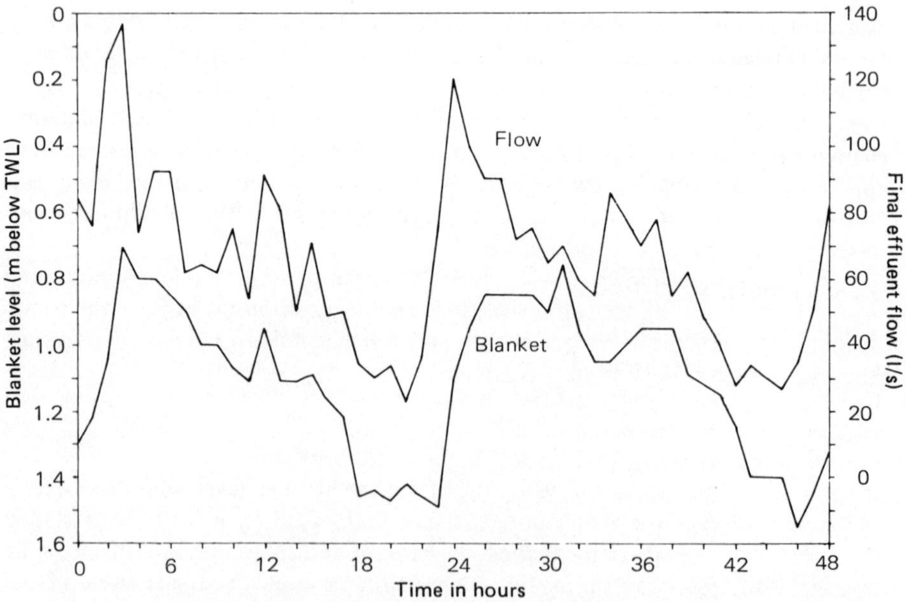

Fig. 13.10 – Variation in sludge blanket level in the final settlement tank with flow.

13.4 CONCLUSIONS

- Alterations to the feed arrangements of the Carrousel aeration tank have considerable influence on sludge settleability. The best settling sludges were obtained with single point feed into an anoxic zone. If designers provide a degree of flexibility in feed arrangements it should be possible to reduce bulking problems at many other plants.
- SSVI, measured at 3.5 g/l, provides a simple and reliable method of monitoring sludge settleability. It is independent of solids concentration over the range normally encountered in activated-sludge plants and is superior to SVI, reflecting changes in settleability which are not apparent in SVI data.
- Mass flux theory can predict within 20 per cent the solids handling capacity of a settlement tank treating activated sludge. The technique of relating mass flux to SSVI can yield useful operational guidelines for settlement tank control. Maximum useful return rate, peak flow rate and maintainable MLSS levels can all be related to SSVI.
- Automatic sludge blanket level control systems could improve final settlement tank performance even with many bulking sludges. A reliable control system has yet to be proved for long term operation on full-scale plants.

13.5 ACKNOWLEDGEMENTS

The authors would like to thank Dr. M. C. Dart, Director of Scientific Services: Mr. E. C. Reed, Director of Operations: Mr. J. L. Wilkins, Divisional Manager, Cotswold Division, and Mr. J. R. Studholme, Divisional Manager, Lambourn Division, Thames Water Authority, for their encouragement in preparing this chapter. The assistance of the many colleagues involved in the project is greatly appreciated.

13.6 NOMENCLATURE

A	cross-sectional area of sedimentation tank (m^2).
C	concentration of suspended solids (kg/m^3).
C_L	limiting concentration of suspended solids at which the flux is a minimum (kg/m^3).
C_f	concentration of suspended solids in feed (mixed liquor) (kg/m^3).
C_u	concentration of suspended solids in underflow (returned sludge) (kg/m^3).
F	mass flux (solids loading) (kg/m^2h).
F_L	limiting solids loading; value of F at minimum of F versus C curve (kg/m^2h).

k	constant, (slope of line log V versus C).
Q_o	volumetric flow rate of waste water (m^3/h).
Q_u	volumetric flow rate of underflow from sedimentation tank (m^3/h).
r^2	coefficient of correlation.
V	hindered settling velocity of suspension (m/h).
V_f	settling velocity of MLSS (m/h).
V_o	constant, (intercept of line log V versus C when C = 0).

13.7 REFERENCES

[1] White, M. J. D. Design and control of secondary settlement tanks. *Water Pollution Control,* 1976, **75**, 459-467.

[2] White, M. J. D. *Settling of Activated Sludge.* Technical Report TR 11, Water Research Centre, 1975, 33 pp.

[3] Johnstone, D. W. M., Rachwal, A. J. and Hanbury, M. J. Settlement characteristics and settlement tank performance in the Carrousel system. *Water Pollution Control,* 1979, **78**, 337-356.

[4] Johnstone, D. W. M., Rachwal, A. J., Hanbury, M. J. and Critchard, D. J. Design and operation of final settlement tanks: Use of stirred specific volume index and mass flux theory. *Paper presented to 33rd International Conference CEBEDEAU,* Liege, Belgium, May 1980.

[5] Tomlinson, E. J. *Bulking – A Survey of Activated Sludge Plants.* Technical Report TR 35, Water Research Centre, 1976, 35 pp.

[6] Chudoba, J., Ottová, V. and Maděra, V. Control of activated sludge filamentous bulking – I Effect of the hydraulic regime or degree of mixing in an aeration tank. *Water Research,* 1973, **7**, 1163-1182.

[7] Waller, C. B. and Hurley, B. E. J. Some experiences in the control of bulking activated sludge. Chapter 12 of this book.

[8] Handley, J. Sedimentation: an introduction to solids flux theory *Water Pollution Control,* 1974, **73**, 230-240.

[9] Dick, R. I. Folklore in the design of final settling tanks. *Journal of the Water Pollution Control Federation,* 1976, **48**, 633-644.

[10] Hibberd, R. L. and Jones, W. F. The design and operation of activated sludge final settling tanks. *Water Pollution Control,* 1974, **73**, 14-32.

Discussion

Chairman: **Professor P. Grau** (Institute of Chemical Technology, Prague, Czechoslovakia)

The discussion was opened by **Dr. M. J. D. White** (Water Research Centre) who supported the authors' views on the shortcomings of Sludge Volume Index (SVI) as a measure of the settling characteristics of activated-sludge. He said that a major disadvantage of SVI was that it varied in an inconsistent way with the initial concentration of suspended solids. Dr. White added that it was common European practice to use a diluted SVI test which overcame such disadvantages to a certain extent but as yet no-one had published a mass flux theory based on this test which allowed satisfactory prediction of the solids loading on a sedimentation tank.

Dr. White also mentioned that the standard cylinder used for measuring SSVI was not manufactured by the WRC or marketed by them. (Editors' note: the standard settling apparatus is available from Triton Electronics Ltd, Bigods Hall, Dunmow, Essex, England.)

Professor D. Jenkins (University of California, Berkeley, USA) congratulated the authors on their paper (Chapter 13) and went on to describe some observations he had made on the behaviour of sludge blankets in final sedimentation tanks. He had often noted that the sludge blanket consisted of more than one layer, the solids concentration being high at the bottom and low within the top layer.

Mr. A. J. Rachwal said that during their work they had always observed a distinct interface between the solids and the clarified effluent in a final sedimentation tank. He considered that there was a gradient of solids concentration within a tank and this increased from essentially the MLSS concentration at the interface to the recycled sludge suspended solids concentration at the bottom of the tank.

Dr. Johnstone and **Mr. Rachwal** both commented on the improvement in effluent quality that resulted from an increase in the sludge blanket depth. They emphasized the effect of the double-sided weir installed in the sedimentation tank on the mechanism of solids loss from the tank at high solids loadings. It was apparent that the majority of the solids were scoured from the sludge blanket and then carried over the outer annulus of the weir only. This scouring effect caused appreciable solids loss even when the sludge blanket was at a considerable depth below the water surface and hence Dr. Johnstone and Mr. Rachwal did not recommend the use of double-sided weirs in activated-sludge final sedimentation tanks.

Mr. G. McBrearty (Thames Water Authority, UK) said that Dr. Johnstone and Mr. Rachwal had mentioned that an activated-sludge sedimentation tank performed the function of a clarifier and a thickener and he asked the authors if they could make some further distinction between these two functions.

Dr. Johnstone replied that, in simple terms, a clarifier was designed to produce an effluent free from suspended solids while a thickener should produce a continuous underflow of thickened sludge. He added that in practice the required surface area of a final tank was determined by the need to transmit the applied solids to the bottom of the tank. Thus for design purposes the sedimentation tank could be considered as a thickener.

Mr. C. B. Waller (Thames Water Authority, UK) said that during the conference the delegates had heard a great deal about filamentous organisms. He asked the authors if they considered that the high quality effluents produced when sludge blankets were high were due to superior substrate removal rates achieved by the population of filamentous micro-organisms. **Mr. Rachwal** replied that the presence of moderate numbers of filamentous organisms did result in a high sludge blanket and providing the sedimentation tank capacity was adequate, high quality effluents were usually produced. However, he considered that the effluent quality was purely a result of physical filtration through the deep sludge blanket and was not due to a more efficient substrate removal achieved by the filamentous organisms.

Mr. G. F. G. Clough (Howard Humphreys and Partners, Leatherhead, Surrey, UK) said that the efficient operation of a sedimentation tank required that the tank be constructed with an adequate depth. This would enable the sludge blanket to rise during peak flows without solids loss occurring. Mr. Clough went on to say that he considered that this was an important Conference but he had heard very little mention of costs, both capital and operating, for some of the bulking control measures put forward. He also said that it was important to consider the prevention of bulking in terms of the entire treatment works' operation. Thus it was no use curing the bulking problem if some other operation in the treatment sequence was upset as a result.

Appendix I – Supplementary Contributions

Supplementary Contribution (i)

QUANTITATIVE FILAMENT COUNTING – A QUICK, SIMPLE METHOD FOR PREDICTION AND MONITORING OF FILAMENTOUS BULKING

A. P. Walker, North West Water Authority

INTRODUCTION

Although there is an abundance of literature which describes and discusses the association of filamentous micro-organisms with bulking sludge, there have been few attempts to quantify filaments objectively so that a numerical value can be compared with Sludge Volume Index (SVI) or Stirred Sludge Volume Index (SSVI). Many workers have used systems of subjective assessment such as the '+ to +++' scale of Rensink [1], the 'slight, moderate or excessive' scale of Farquar and Boyle [2], or the arbitrary 10-point scale of Forster and Dallas-Newton [3]. Chudoba *et al.* [4] estimated the relative proportions of the microscope field occupied by filamentous and zoogloeal bacteria, and Raleigh *et al.* [5] used a system of photomicrographs and subjective assessment.

Pipes [6] went some way towards a true numerical count by counting the total number of actual filaments but took no account of individual filament length. Finstein and Heukelekian [7] came closer by measuring the total length of filaments protruding through the floc, and Sezgin and Jenkins [8] actually measured the total filament length per unit weight of dry floc solids. Unfortunately, the techniques described in both of the latter papers were lengthy and do not lend themselves to the routine laboratory situation.

The objective of the method described here was therefore to provide a quick and simple counting technique capable of removing the element of 'guesstimation' from filament determination sufficiently to provide a practical tool for use in monitoring and prediction of filamentous bulking. It was also intended that the method should lend itself to both centralized laboratory facilities remote from the treatment plant and to the on-site laboratory.

METHODOLOGY

Samples of at least 50 ml of mixed liquor are taken to the laboratory with the minimum of delay. Each sample is then shaken thoroughly and a small quantity is drawn up into a teated pipette and run under the coverglass of a calibrated Lund nanoplankton counting chamber (these 'Lund cells' were obtained from the Water Research Centre, Medmenham Laboratory). The liquor is run in at the centre of one end of the cell to minimize uneven distribution due to wall effects. The cell is then examined under a microscope fitted with a Whipple eyepiece graticule (graticule obtained from Gallenkamp), using the 10× objective or nearest equivalent. The graticule superimposes a grid of 100 squares on to the image of the activated sludge (see Fig. 1a). The number of filaments crossing each of the ten squares on each diagonal is counted. Only those filaments which are within a square for at least 0.6 of one square width are counted. If two filaments each cross a square for less than this minimum distance they are counted as one filament unit. The counts derived from Fig. 1a are given in Fig. 1b.

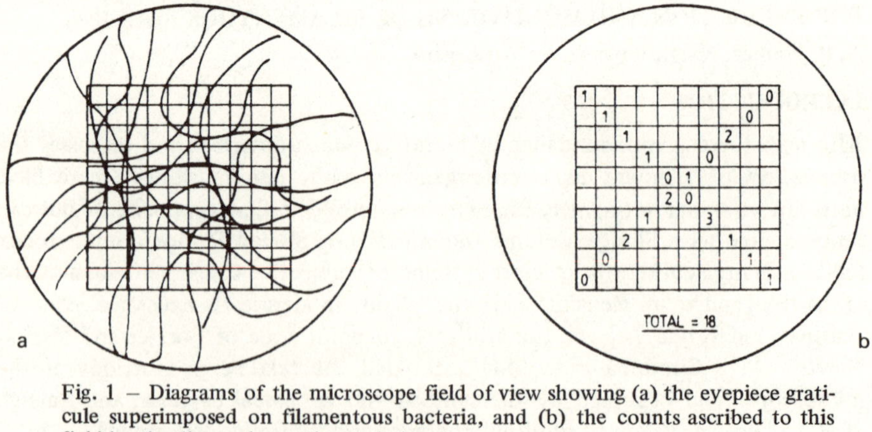

Fig. 1 – Diagrams of the microscope field of view showing (a) the eyepiece graticule superimposed on filamentous bacteria, and (b) the counts ascribed to this field by the diagonal counting technique.

In practice it is essential to count one square at a time, racking the microscope through the full depth of focus in each square, since the depth of the Lund cell is considerably greater than the depth of focus at this magnification. Five fields are normally counted, evenly spaced along the axis of the cell, thus giving a total count over 100 squares.

Provided that a stage micrometer slide has been used to measure the size of the grid squares, and that the cell has been calibrated such that the liquid depth is known, it is then possible to calculate the length of filament per ml of mixed liquor by making the assumption that the length of filament crossing a square is on average equivalent to the width of that square. The minimum filament traverse of 0.6 square width was therefore chosen to balance those filaments which cross a square diagonally (i.e., 1.4 × square width). If the Mixed-Liquor Suspended Solids (MLSS) is determined on the same sample, the length of filament per unit weight of dry solids can be calculated as follows:

CALCULATION OF RESULTS

$$\text{Filaments (cm/mg MLSS)} = \frac{N \times F \times 1000}{\text{MLSS (mg/l)}}$$

where N = count of filaments per 100 squares

$$F = \frac{\text{Width of square (cm)}}{\text{area of 100 squares (cm}^2\text{)} \times \text{cell depth (cm)}}$$

F is a constant for any specific microscope/cell combination at fixed magnification

CALIBRATION OF LUND CELL

Place the coverglass on the cell and measure the length and width of the enclosed volume, then weigh the cell + coverglass to four decimal places. Run distilled water into the cell with a teated pipette until just full. There must be no bubble trapped, nor must there be any tendency of the coverslip to 'lift'. Reweigh the cell. Assuming a density of 1.0 g/ml for water, the difference in weight is equivalent to the total cell volume.

$$\text{cell depth (cm)} = \frac{\text{difference in weight (g)}}{\text{area under coverglass (cm}^2\text{)}}$$

A summary of the counting procedure is shown diagrammatically in Fig. 2.

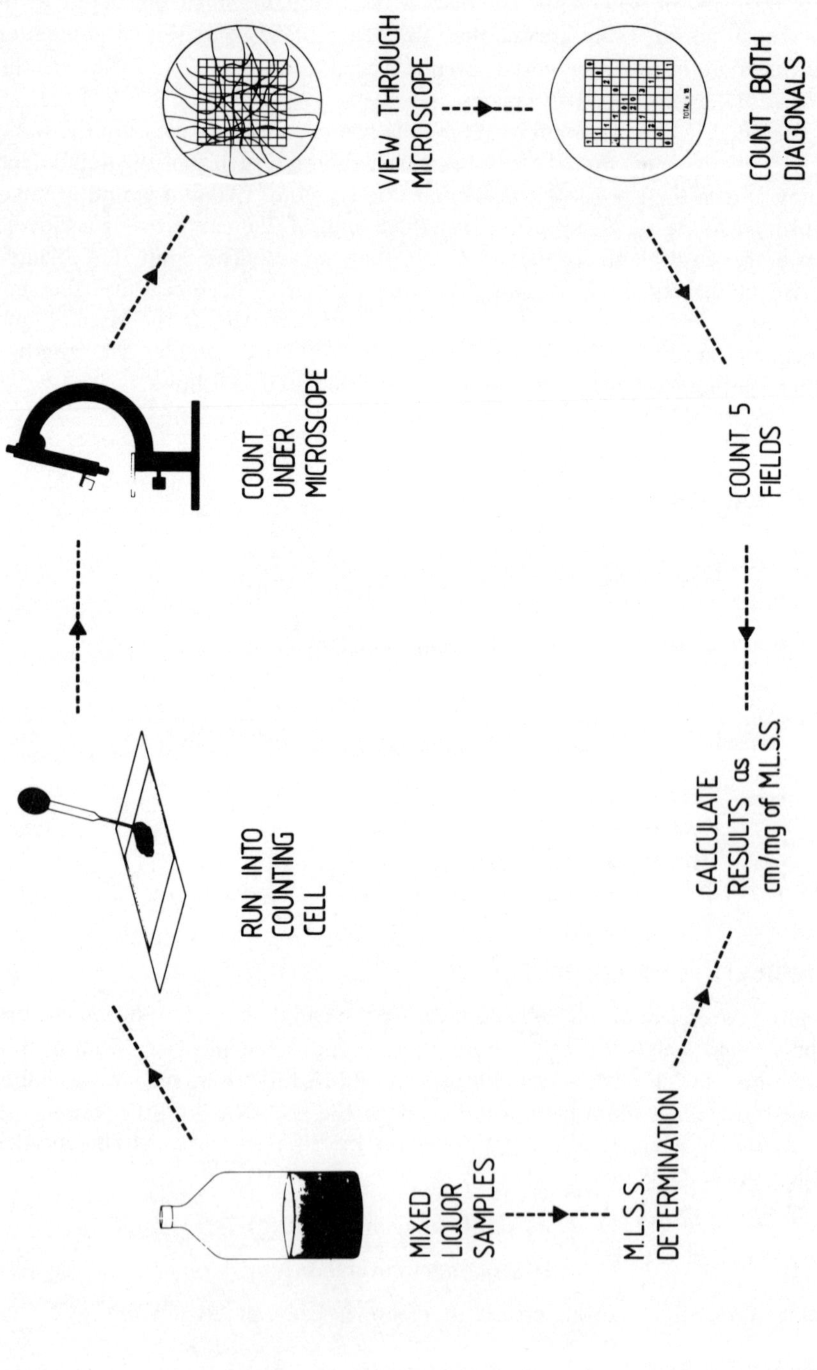

Fig. 2 – Summary of counting procedure.

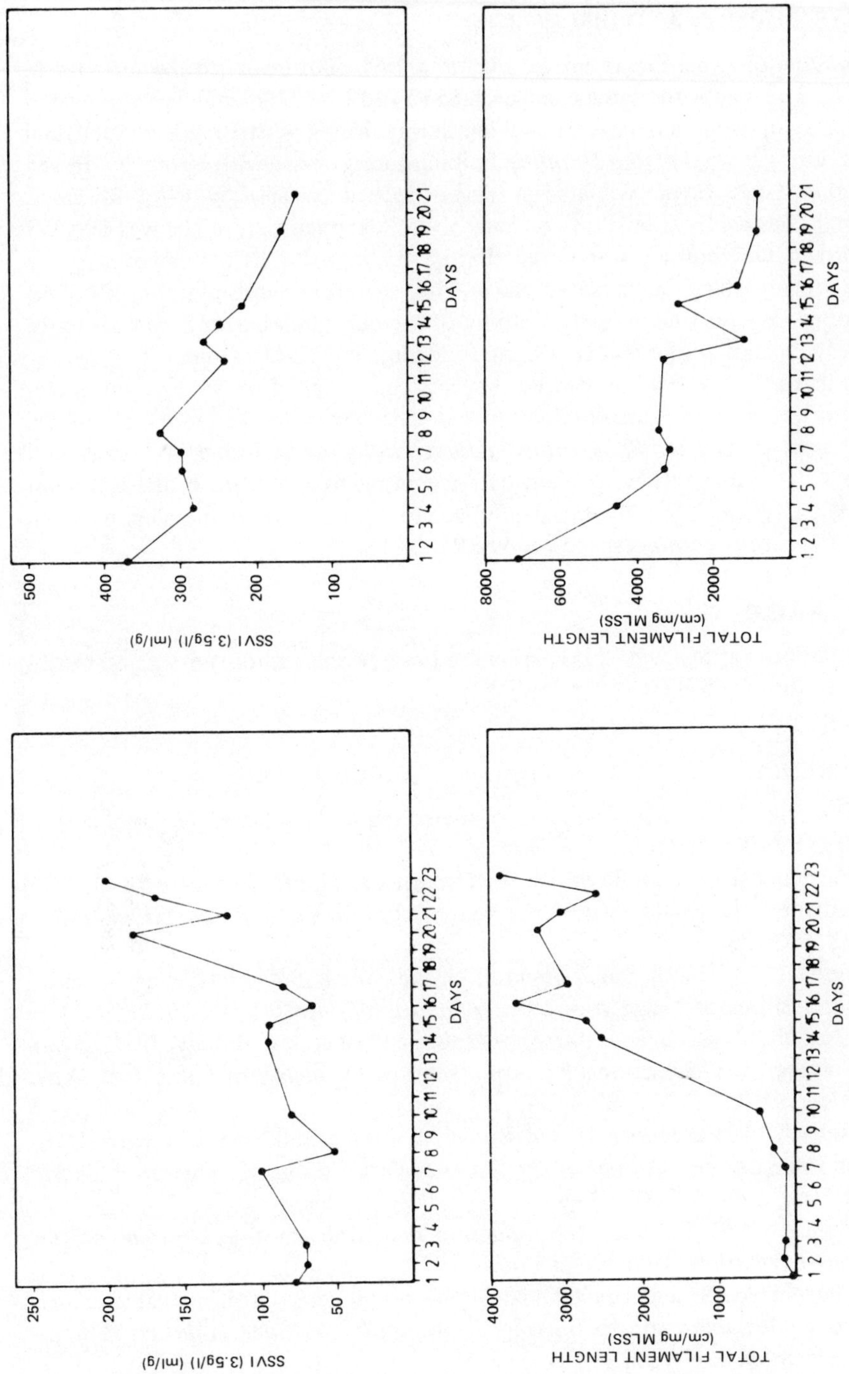

(a) During development of bulking due to *Thiothrix*

(b) During recovery from *Thiothrix* bulking involving hydrogen peroxide dosing from day 1 to day 8.

Fig. 3 – Comparison of total filament length per milligram of MLSS with SSVI.

USE OF METHOD AND DISCUSSION

The results of experiments carried out on a $1m^3$ pilot plant are given in Figs 3a and 3b. The graphs show the changes in SSVI and total filament length associated with development of *Thiothrix* bulking during dosing with 5 mg/l sulphide, and again during recovery from *Thiothrix* bulking during dosing with 30 mg/l hydrogen peroxide. (Dose rates calculated as mg/l of settled sewage flow.) In both cases, marked changes in total filament length were detected before changes in SSVI confirmed that bulking was developing or abating, the discrepancy during the development stage being seven days. This therefore indicated that filament counting may provide an early warning of incipient bulking, and may also give an early indication of a recovery trend following remedial treatment.

Filament counting has also been successfully applied to the prediction and monitoring of filamentous bulking in full-scale plants dominated by *Thiothrix, Microthrix parvicella, Sphaerotilus natans, Haliscomenobacter hydrossis* and Type 1701. Filaments were identified according to the scheme of Eikelboom [9]. Work is being continued to determine more closely the relationships between the different filamentous species and SSVI.

DISCLAIMER

The opinions expressed in this chapter are those of the author and not necessarily those of the North West Water Authority.

REFERENCES

[1] Rensink, J. H. New approach to preventing bulking sludge. *Journal of the Water Pollution Control Federation,* 1974, **46**, 1888–1894.

[2] Farquar, G. J. and Boyle, W. C. Occurrence of filamentous organisms in activated sludge. *Journal of the Water Pollution Control Federation,* 1971, **43**, 603–620.

[3] Forster, C. F. and Dallas-Newton, J. Activated sludge settlement – some suppositions and suggestions. *Water Pollution Control,* 1980, **79**, 338–351.

[4] Chudoba, J., Grau, P. and Ottova, V. Control of activated sludge filamentous bulking – II. Selection of micro-organisms by means of a selector. *Water Research,* 1973, **7**, 1389–1406.

[5] Raleigh, C., Caropreso, F. and Brown, J. Attack bulking sludge with H_2O_2 and microscope. *Journal of the Water Pollution Control Federation, Deeds & Data,* 1974, D1, D7–D8.

[6] Pipes, W. O. Bulking, deflocculation and pinpoint floc. *Journal of the Water Pollution Control Federation,* 1979, **51**, 62–70.

[7] Finstein, M. S. and Heukelekian, H. Gross dimensions of activated sludge flocs with reference to bulking. *Journal of the Water Pollution Control Federation,* 1967, **39**, 33–40.

[8] Sezgin, M., Jenkins, D. and Parker D. S. A unified theory of filamentous activated sludge bulking. *Journal of the Water Pollution Control Federation,* 1978, **50**, 352-381.
[9] Eikelboom, D. H. Filamentous organisms observed in activated sludge. *Water Research,* 1974, **9**, 365-388.

Supplementary Contribution (ii)

SURVEY OF INDUSTRIAL ACTIVATED SLUDGES: FLOC MORPHOLOGY AND ISOLATION OF MICRO-ORGANISMS

J. P. Salanitro and **W. E. Rader**, Shell Development Company, Houston, Texas, USA.

INTRODUCTION

The isolation and identification of the more numerous bacterial species in sludge flocs are useful parameters for studying the microbial ecology of activated sludges. Analyses of the distribution of filamentous and non-filamentous populations in mixed liquor suspended solids are important in the assessment of:
(i) the morphology and physiology of flocs;
(ii) the occurrence and predisposition of sludges to bulk, foam or grow as dispersed and pinpoint floc suspensions;
(iii) the settleability of the sludge (flocculation and dewatering strategies);
(iv) the treatability of various chemical wastes.

Studies on microbial composition of municipal sewage sludges from Europe and the US have indicated that flocs may contain one or more of at least 20 different types of filamentous species [1-4]. Predominant filamentous organisms identified include *Nocardia, Haliscomenobacter, Sphaerotilus, Microthrix,* and several unnamed types. Some of the more common non-filamentous bacteria are *Pseudomonas, Zoogloea, Arthrobacter, Flavobacterium* and other gram-negative bacilli and gram-positive cocci [5]. There is little information, however, on the microbial structure and composition of industrial activated sludges, especially from petrochemical manufacturing plants. In this chapter some preliminary studies on the sludge floc structure and types of microbes isolated from refinery and chemical plant mixed liquor suspended solids are described. The results indicate that these industrial sludges have a filamentous or non-filamentous structure and contain organisms which secrete a fibrillar-like material. This extracellular biopolymer may in part be responsible for floc rigidity and intercellular bridging in the microbial aggregate.

EXPERIMENTAL PROCEDURES

Petrochemical plant sludges

Activated-sludge samples were obtained from the aeration basins of chemical industry or refinery treatment plants. The types of products manufactured, and waste water treatment operations of the plants surveyed in this study are given in Table 1. Samples obtained from distant plants were preserved in ice (24 hours) until processed for microscopy and plating on media.

Table 1. – Microbial composition of sludges from petrochemical waste water treament plants

Plant	Products Manufactured	Method of Waste Treatment	Floc Morphology	Predominant Organisms Isolated*
A	alcohols, ketones, olefins, epoxy resins, phenol, and plastic monomers	activated sludge	filamentous	*Nocardia* *Sphaerotilus* *Arthrobacter* *Pseudomonas* *Zoogloea*
B	nonionic surfactants	activated sludge	filamentous	*Nocardia* *Arthrobacter* *Pseudomonas*
C	synthetic rubbers	biodisk	filamentous	*Nocardia* *Pseudomonas* *Arthrobacters*
D	C_3 alcohols and ketones, vinyl chloride	UNOX®	non-filamentous	*Arthrobacter* *Pseudomonas* *Zoogloea*
E	herbicides and insecticides	activated sludge	non-filamentous	*Arthrobacter* *Pseudomonas*
F	gasoline, oil, and anionic surfactants (refinery)	activated sludge	non-filamentous	*Arthrobacter* *Pseudomonas*

* Tentatively identified from isolates of sludges recovered on TGEVA medium.

Culture techniques

Sludge samples (10 to 15 ml) were exposed to ultrasonic vibration (Heat Systems Sonifier) for 30 seconds at 125 watts. Previous work in our laboratory and by others [6] showed that much higher colony counts were obtained when flocs were disrupted by sonication for short periods. Disrupted flocs were quickly diluted 10^7-fold in filter sterilized clarifier effluent or a sterile mineral salts solution (see sludge extract media below). One ml from each of the 10^5 to 10^7-fold dilutions was added (in triplicate) to sterile petri plates, molten (50 °C) agar media added and mixed to distribute the inoculum. Plates were incubated at 30 °C and colony counts made after 10 to 14 days.

Culture media

The following agar media were compared for recovery of viable cells:

(a) nutrient agar (Difco);
(b) nutrient + 2.0 per cent NaCl;
(c) casitone-glycerol-yeast extract; CGY [7];
(d) peptone-glucose-yeast extract; PGY;
(e) tryptone-glucose-meat and yeast extract-vitamins; TGEVA [8];
(f) sludge extract.

Sludge extracts (from plant A and Houston municipal sanitary district) were prepared from autoclaved aeration basin mixed liquor suspended solids (30 minutes at 15 psi), cooled and centrifuged (10 000 rev/min, 10 minutes) to remove cells and debris. The supernatant extracts were adjusted to pH 7.5 and stored at 4 °C. Sludge extract agar media contained extracts at 10 to 40 per cent v/v, supplemented with a mineral solution of the following salts to the final concentration:

	mg/l
NH_4Cl	383
NaCl	20 000
$Na_3PO_4.12H_2O$	308
KCl	191
$MgSO_4.7H_2O$	152
$FeCl_3.6H_2O$	5
$ZnSO_4.7H_2O$	2.2
$MnCl_2.4H_2O$	1.8
$CuSO_4.5H_2O$	0.17

The agar (1.5 per cent w/v) and other components used in the preparation of media were adjusted to a pH value of 7.5 and autoclaved in screw cap bottles for 30 minutes at 15 psi.

Identification of isolated bacteria

Bacterial strains were isolated from TGEVA media for all plant sludges examined. All colonies on a plate containing 50 to 150 colonies were picked to TGEVA slants and then subcultured to TGEV liquid medium for 24 to 72 hours or until good growth occurred. Filamentous and non-filamentous bacterial cultures were identified by morphological and physiological features according to established methods in the literature [1-3,8] and Bergey's Manual of Determinative Bacteriology (8th edition), 1974.

Microscopic observations

Sludge flocs were examined under dark-field phase optics with a Zeiss photomicroscope III and photographs were taken at 110× magnification. Pure cultures of isolated bacteria grown in TGEV medium and sludge flocs were also prepared for scanning electron microscopy (SEM) by primary fixation in 2 per cent glutaraldehyde – 0.1M cacodylate buffer pH 7.0 (three hours at 4 °C) and post fixed in 2 per cent OsO_4 – 0.1M collidine buffer pH 7.0 (1 hour). Specimens were then dehydrated through a graded series (50 to 100 per cent) of ethanol, critical point dried with liquid CO_2, mounted on stubs and coated with gold-palladium dust to 100–200Å thickness. The bacteria and flocs were viewed and photographed in a JEOL 100CX scanning/transmission electron microscope operated at 20KeV.

Table 2. – Comparison of Various Culture Media for Enumeration and Isolation of the Predominant Bacteria in Sludge Flocs[a)]

Medium	Recovery[b)] (%)
Nutrient Agar	21
Nutrient Agar + 2.0 per cent NaCl	27
Plate Count Agar	25
Casitone-glycerol-yeast extract	30
Peptone-glucose-yeast extract	35
TGEVA	37
Sludge Extract (10 per cent v/v)	40
Municipal Sludge Extract (10 per cent v/v)	40

a) Sludge samples (plant A in Table 2) were treated by sonic disruption of 30 seconds, diluted and plated on media. Plates were incubated at 30 °C for 10 to 14 days and colonies counted.
b) Colony counts/microscopic count × 100 per cent. Microscopic counts in samples varied from $3–6 \times 10^8$ bacteria/ml sludge.

RESULTS AND DISCUSSION

Comparison of culture media

Several culture media were evaluated for enumerating and isolating the predominant bacteria in sludge flocs (Table 2). Sludge from chemical plant A Table 1 was used in comparing the efficiency of plate count recoveries with the different media. About 30 to 40 per cent of the microbial cell population (direct microscopic count) in disrupted flocs could be cultured in all media except common bacteriological media. Only 20 to 25 per cent of the microscopic count could be grown on nutrient agar and plate count agar. It is not certain from these viable cell recoveries whether a large fraction (60 to 70 per cent) of the bacteria in these sludges are dead cells or organisms unable to grow on the available nutrients in these media. Adding a B-vitamin mixture, or small amounts of chemical plant waste water to sludge extract media, or increasing the sludge concentration to 40 per cent v/v, did not improve viable count recoveries. In general, colony types on TGEVA were more diverse and larger than those recovered on sludge extract agar. TGEVA was used, therefore, as the culture medium to enumerate and isolate the microbes in the survey of several petrochemical plant sludges.

Survey of petrochemical plant sludges, floc structure, and microbial composition
Table 1 lists the different activated sludges analysed according to plant, method of treatment, floc structure and bacteria isolated. Three chemical plant sludges (A, B, and C) contained highly filamentous flocs while those from two chemical plants (D and E) and a refinery (F) were composed of non-filamentous bacteria. Figure 1 contains photographs illustrating the filamentous or non-filamentous nature of the sludges as viewed under dark-field phase microscopy. A and B flocs appear to be less dense with numerous exposed bacterial filaments. These sludges settle poorly (SVI greater than or equal to 150 ml/g) compared to the more densely aggregated flocs of C shown in Fig. 1. In contrast, samples of D, E, and F sludges show a floc structure which is highly fragmented and essentially devoid of filamentous organisms. In addition, D and F samples have numerous pinpoint flocs and dispersed bacteria. E flocs occasionally contain filamentous-type species similar to cyanobacteria. The predominant filamentous organisms isolated from filament-bearing sludges on TGEVA medium include species of *Nocardia* and *Sphaerotilus* while the major non-filamentous microbes recovered were those of *Pseudomonas, Arthrobacter,* and *Zoogloea. Pseudomonas* and *Arthrobacter* were also the predominant bacteria isolated from non-filamentous flocs (Table 1).

Scanning electron micrographs (SEM) of the sludge flocs examined are shown in Fig. 2. These photographs illustrate the extensive network of filamentous species in A, B, and C flocs compared with D, E, and F. SEM of A and F flocs (Fig. 2, B and F) show the extensive fibril-like material interconnecting the microbes in the floc matrix. Examination of pure cultures of isolated strains of *Nocardia, Pseudomonas,* and *Zoogloea* (Fig. 3) under SEM also indicates that these species produce an exocellular fibril polymer similar to the material seen in the SEM of whole sludge flocs. These results suggest that biopolymers are involved in bridging filamentous and non-filamentous bacteria and may impart rigidity to the sludge floc structure.

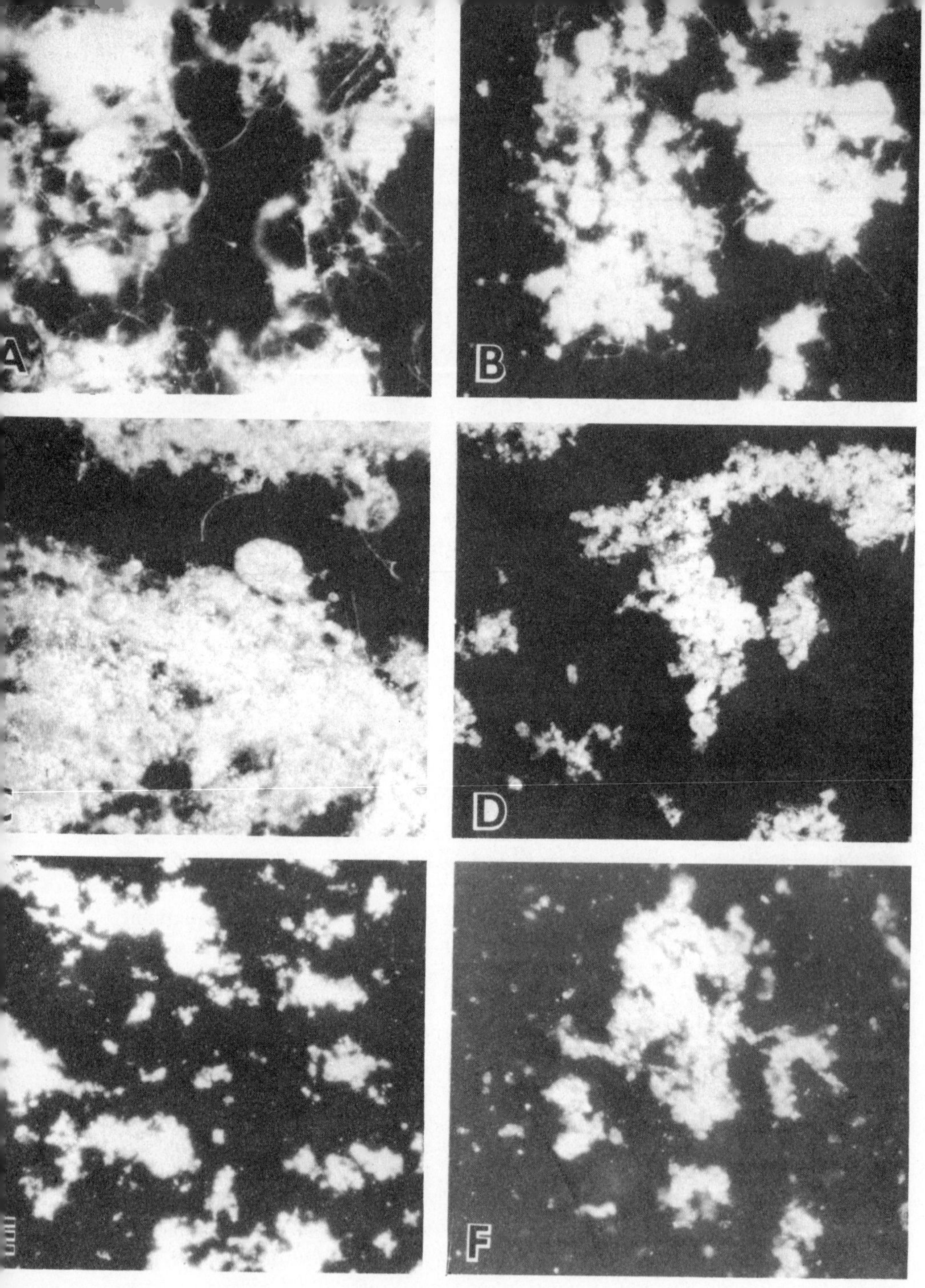

Fig. 1 – Dark-field photomicrographs of sludge flocs (160×) from treatment plants A, B, C, D, E and F (designated as in Table 1).

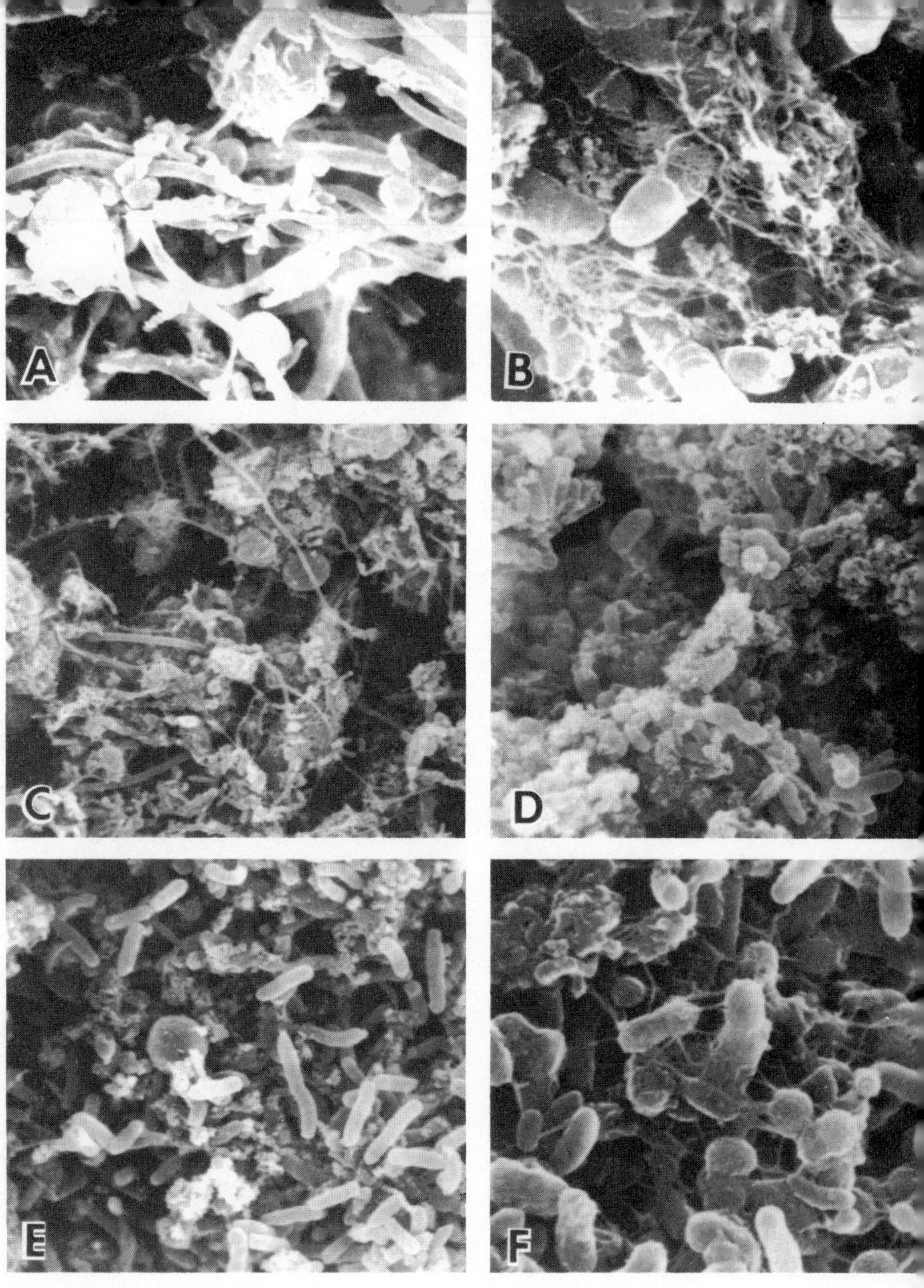

Fig. 2 – Scanning electron micrographs of sludge flocs from treatment plants A, B, C, D, E, and F: A (plant A, 5000×); B (plant A, 20 000×); C (plant B, 1000×); D (plant D 5000×); E (plant E, 5000×); F (plant F, 20 000×).

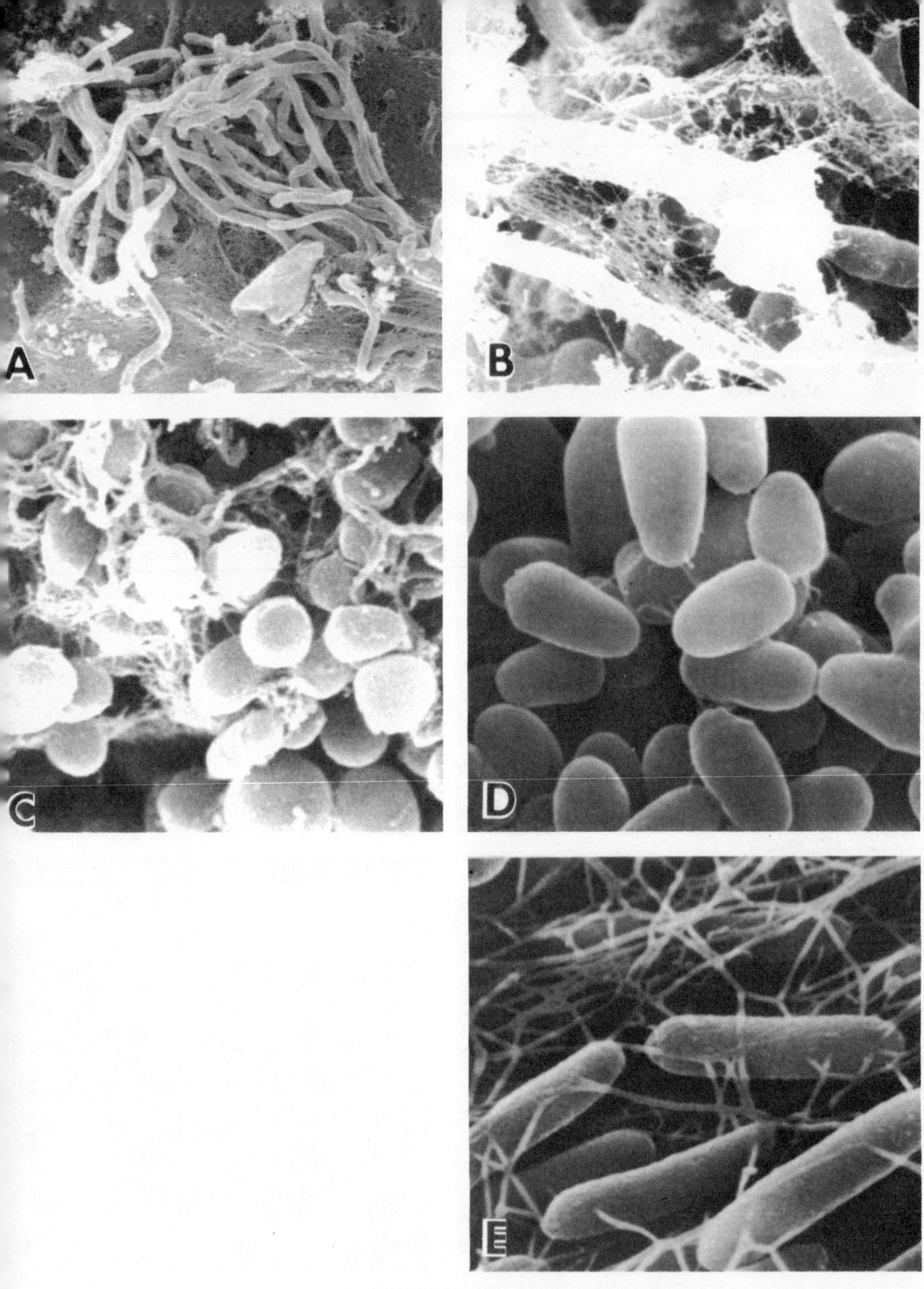

Fig. 3 – Scanning electron micrographs of predominant bacterial strains isolated from sludges. All isolates were cultured on TGEVA or sludge extract media: A, *Nocardia* (20 000×); B, *Sphaerotilus* (5000×); C, *Pseudomonas* (20 000×); D, *Arthrobacter* (20 000×); E, *Zoogloea* (20 000×).

REFERENCES

[1] Eikelboom, D. H. Filamentous organisms in activated sludge. *Water Research,* 1975, **9**, 365–388.

[2] Eikelboom, D. H. Identification of filamentous organisms in bulking activated sludge. *Progress in Water Technology,* 1977, **8**, 153–161.

[3] van Veen, W. L. Bacteriology of activated sludge, in particular the filamentous bacteria. *Antonie van Leeuwenhoek,* 1973, **39**, 189–205.

[4] Strom, P. F. and Jenkins, D. *Filamentous Micro-organisms in Activated Sludge Plants of the United States, Particularly California.* Progress Report Presented at the 51st Annual Conference of the California Water Pollution Control Association. April 26, 1979.

[5] Pike, E. B. and Curds, C. R. The microbial ecology of the activated sludge process. In *Microbial Aspects of Pollution,* Sykes, G. and Skinner, F. A. (eds.), 1971, Academic Press, New York.

[6] Banks, C. J. and Walker, I. Sonication of activated sludge flocs and the recovery of their bacteria on solid media. *Journal of General Microbiology,* 1977, **98**, 363–368.

[7] Pike, E. B., Carrington, E. G. and Ashburner, P. A. An evaluation of procedures for renumerating bacteria in activated sludge. *Journal of Applied Bacteriology,* 1972, **35**, 309–321.

[8] Lighthart, B. and Oglesby, R. T. Bacteriology of an activated sludge wastewater treatment plant – A guide to methodology. *Journal of the Water Pollution Control Federation,* 1969, **41**, 267–281.

Supplementary Contribution (iii)

USE OF AN ANOXIC ZONE TO IMPROVE ACTIVATED SLUDGE SETTLEABILITY

G. J. Price, Wessex Water Authority

INTRODUCTION

A significant improvement in Stirred Specific Volume Index (SSVI) (at 3.5 g/l), from 155 to 112 ml/g was achieved by introducing an anoxic zone in front of the existing four aeration pockets at Thingley Sewage Treatment Works (STW). Thingley STW is a works with a population equivalent of 10 900, an average flow of 3600 m^3/d and Dry Weather Flow of 2400 m^3/d. During early 1978 extensions to the works were completed and the aeration capacity increased from four to six pockets, each pocket having a capacity of 270 m^3. Since a fully nitrified effluent could be obtained with the existing four pockets it was decided to try and improve the SSVI by using one of the extra pockets as an anoxic zone, i.e. 20 per cent of the total aeration capacity, with the aerator switched off. In order to ensure that the sludge did not accumulate in the bottom of the aeration pocket, the aerator was switched on for a two-hour period twice a day. The retention time in the anoxic zone is approximately one hour for the settled sewage and return sludge flow. The activated sludge loading was kept similar to the previous period by reducing the MLSS from 3300 to 2800 mg/l.

The results are shown in Table 1. There was an immediate improvement in settleability after the anoxic zone was installed, with a long term significant reduction in SSVI (at 3.5 g/l) from 155 to 112 ml/g. The final effluent was still fully nitrified and there was an increase in the removal of nitrogen from the incoming feed sewage.

Table 1 – SSVI (at 3.5 g/l) before and after installation of anoxic zone

Period	MLSS (mg/l)	SSVI (at 3.5 g/l) (ml/g)	Primary Tank effluent Amm. N (mg/l)	Final effluent Amm.N (mg/l)	Final effluent TON (mg/l)	Final effluent BOD (mg/l)	Final effluent SS (mg/l)
July 77 to Mar 78	3300	155	33.2	1.3	10.7	11	13
May 78 to Nov 78	2800	112	39.0	1.1	10.1	11	14

Since November 1978 two anoxic zones in series have been tried, i.e. 40 per cent of the total aeration capacity, on two occasions but with no further improvement in SSVI (see Table 2). In January 1981 the sixth pocket was brought into use, hence the anoxic zone was 17 per cent of the total aeration capacity again with a retention time of one hour. The settleability remained unchanged.

Table 2 – Variation in SSVI (at 3.5 g/l) with different modes of operation

Period	Mode of Operation	SSVI (at 3.5 g/l) (ml/g)	MLSS (mg/l)
Nov 78 to Mar 79	3 pockets plus 2 anoxic zones	112	2650
Apr 79 to June 79	4 pockets plus 1 anoxic zone	70	2760
July 79 to Aug 79	3 pockets plus 2 anoxic zones	88	3640
Sept 79 to Jan 80	4 pockets plus 1 anoxic zone	93	3200
Jan 80	5 pockets plus 1 anoxic zone	93	3480

Supplementary Contribution (iv)

THE EFFECT OF A CHANGE IN LONGITUDINAL MIXING ON THE SETTLEABILITY OF AN ACTIVATED SLUDGE

M. D. Humphries, Wessex Water Authority

INTRODUCTION

The trial was carried out on Blackwell Hams Sewage Treatment Works which is a conventional activated-sludge plant treating sewage from Chippenham in the Bristol Avon Division of Wessex Water Authority. The specifications of the plant are as follows:

Design Dry Weather Flow – (DWF) 10 450 m^3/day.
Total treatment capacity – 39 000 m^3 /day.
Preliminary treatment – Mechanical bar screen followed by Pista grit trap.
Primary treatment – Two circular sedimentation tanks with half bridge scrapers, total volume of 2400 m^3.
Biological treatment – Simplex surface aeration system operating in two lanes of four pockets each, total volume of 2950 m^3. Dissolved Oxygen controlled automatically by adjustable weir linked to DO probe.
Final settlement – Two circular settlement tanks with half bridge scrapers, total volume of 3250 m^3.
Sludge recycle – Manually controlled bellmouths feeding single archimedian screw, maximum recycle rate 11 000 m^3 /day.

At the time of the trial the DWF was 5000 m^3 /day and the Stirred Specific Volume Index (SSVI) at 3.5 g/l was typically 150 ml/g. Because the plant was operating below its design capacity, the bulky sludge did not cause any problems except during storm conditions when it was found that the plant lost solids from the final tanks before the flow reached maximum design flow. The normal arrangement of the aeration pockets was as shown in Fig. 1.

METHOD AND RESULTS

In an attempt to improve the settlement of the activated sludge it was decided to decrease longitudinal mixing (increase degree of 'plug-flow') by running the two lanes in series instead of in parallel. The hand-stocks between the centre channel and pockets four and five were rearranged, and the weir in pocket four was taken off remote control and locked in the fully raised position, giving the flow pattern shown in Fig. 2.

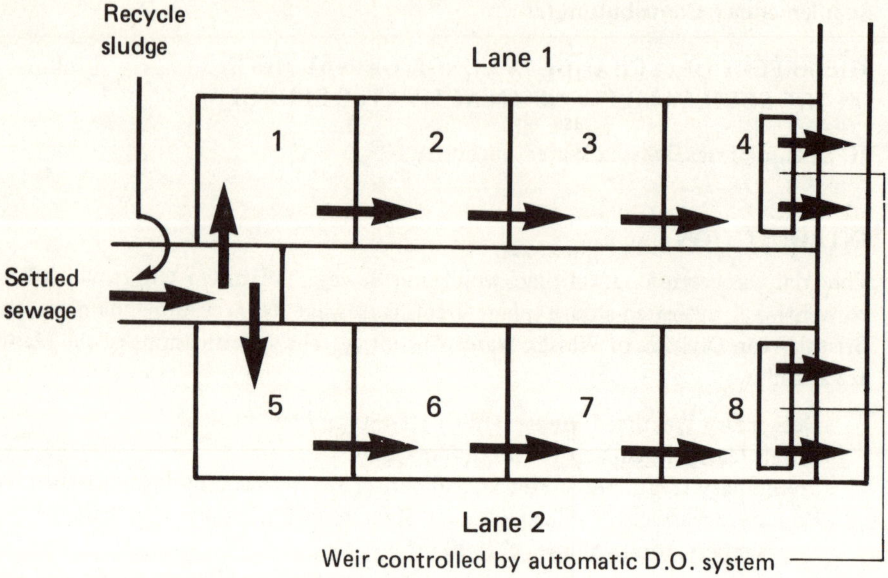

Fig. 1 – Layout of aeration tanks before modification.

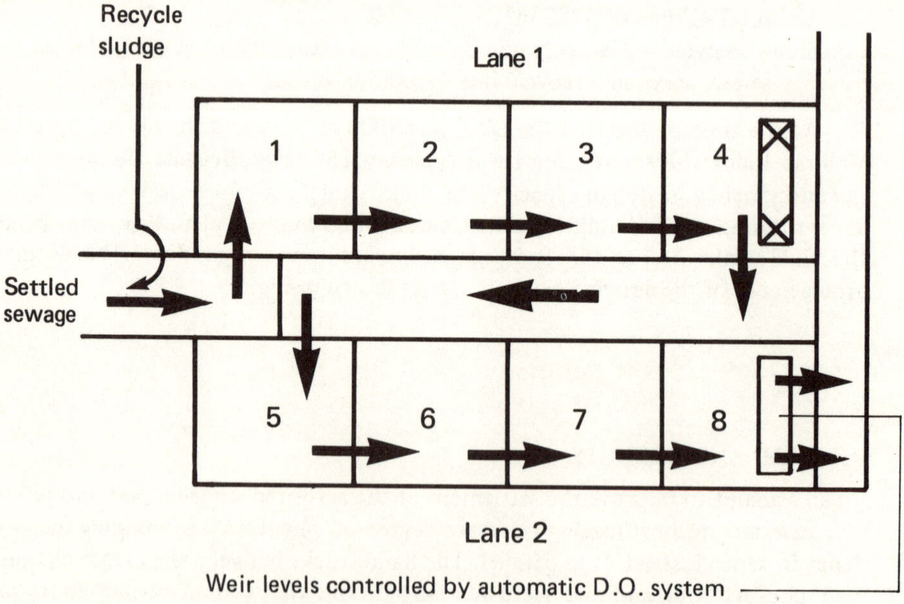

Fig. 2 – Layout of aeration tanks after modification to decrease longitudinal mixing.

The normal weekly monitoring of the Stirred Sludge Volume Index was continued, the results are shown in the form of a graph in Fig. 3, with SSVI (at 3.5 g/l) plotted against time. The graph is divided into four phases, as follows:

Phase I — Control phase showing the SSVI (at 3.5 g/l) for two months before the trial.

Phase II — Rapid drop in SSVI over a period of approximately five sludge ages (sludge age was eleven days).

Phase III — Period of low stable SSVI.

Phase IV — Period of drastic change in loading and flow, caused by the diversion to Blackwell Hams of the flow from a nearby works which was closed down.

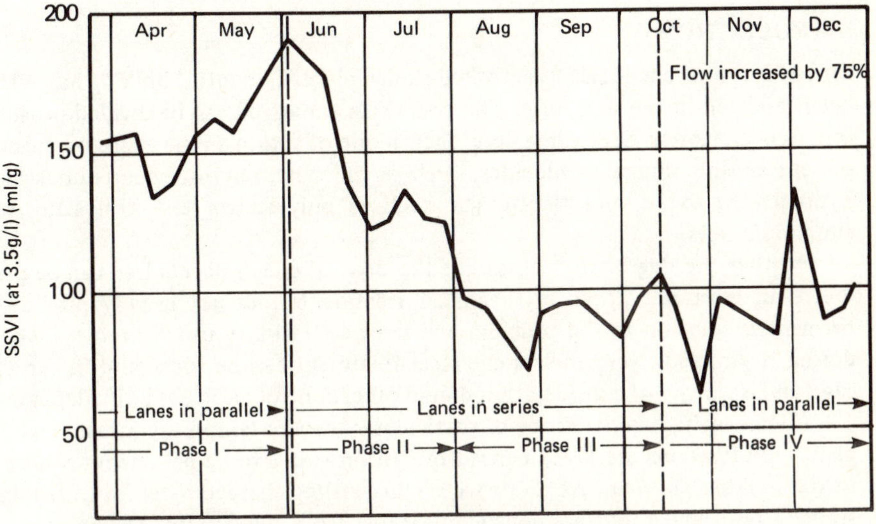

Fig. 3 – Graph showing SSVI (at 3.5 g/l) of Blackwell Hams activated sludge, April to December 1980.

In fact the load to the plant increased by 75 per cent at the beginning of Phase IV, and it was found that the head difference between the first and last pockets was enough to overload the aerator motors in pockets one and two when full immersion was required. For this reason the flow pattern was changed back to two parallel lanes. The SSVI (at 3.5 g/l) has not increased to the former high values, so it would appear that the increased sludge loading has also caused an improvement.

A further experiment is planned to adjust the immersion levels of the aerators individually so that the series flow pattern can be operated at the higher loading. This may result in a further improvement in settleability.

The effluent quality did not vary significantly from the normal high standard during the trial, and nitrification was maintained throughout.

Supplementary Contribution (v)

THE COST OF CHEMICAL TREATMENT TO CONTROL THE BULKING OF ACTIVATED SLUDGE

B. Chambers and **E. J. Tomlinson**, Water Research Centre, Stevenage Laboratory

INTRODUCTION

The addition of chemicals to activated-sludge plants to control bulking has often been reported in the literature. The chemicals employed can be divided broadly into two categories, depending upon their mode of action on the micro-organisms present in the sludge – biocides, such as chlorine, chlorine compounds and hydrogen peroxide, and flocculants, such as polyelectrolytes, iron salts and aluminium salts.

Neither the degree of success nor the dose of chemical required can be predicted in advance. Moreover, chemical treatment does not usually provide a permanent solution to the problem and poor settleability usually returns when dosing is stopped. Nevertheless chemical treatment can be successful in certain cases and is worthy of consideration should other control measures be inadequate.

Chemical treatment can be of particular value as a temporary measure while plant modifications are being carried out to provide a more permanent solution to the bulking problem. At plants where the settling characteristics of the sludge are very poor and solids are continuously lost from the system, certain control measures may be difficult to apply. Thus, in such a situation it can be impossible to lower the sludge loading by increasing the Mixed Liquor Suspended Solids (MLSS) concentration. Temporary dosing with a suitable chemical can often improve the settleability of the sludge to such an extent that other control measures may be employed to result in permanently stable plant operation.

BIOCIDAL SUBSTANCES

Chlorine

Chlorine and chlorine compounds such as sodium hypochlorite are very toxic to micro-organisms in quite small doses. Overdosing with chlorine will inhibit nitrification, kill protozoa, and produce a turbid effluent as a result of cell disintegration and lysis.

Reported effective doses and addition points vary and it is difficult to draw overall conclusions concerning chlorine treatment. An approach which has been recommended involves commencement of dosing at very low concentration into the return sludge. The dose is gradually increased until the required improvement in settleability is achieved. The symptoms of overdosing described above may appear before this point is reached.

A few of the reported effective doses, to be added on a continuous basis, are shown in Table 1.

Table 1 – Effective doses of chlorine to control bulking sludge

Dose		Reference
kg Cl_2/1000 kg return sludge	mg Cl_2/litre sewage	
2.5	-	[1]
0.1 to 1.0	-	[2]
-	10 - 20	[3]

Hydrogen Peroxide

The addition of this chemical to activated-sludge plants has been shown to be effective in a number of cases for controlling bulking sludge. Hydrogen peroxide has the advantage of supplying oxygen to the process, since this is one of the decomposition products of the substance.

Keller and Cole [4] have attempted to quantify the amount of hydrogen peroxide required on a rational basis. Their data are expressed as the mass of hydrogen peroxide per mass of Mixed Liquor Suspended Solids necessary for a 50 per cent reduction in Sludge Volume Index (SVI). Thus 0.1 kg H_2O_2/kg MLSS causes this reduction in seven days while 0.4 kg H_2O_2/kg MLSS acts in about one day. Additions below the former concentration had no effect and doses above the latter had the disadvantage of causing deflocculation and increased effluent turbidity.

FLOCCULATING CHEMICALS

Flocculants employed to control bulking sludge include iron salts, alum and various synthetic polyelectrolytes. There is some doubt as to whether the inorganic salts used in this manner are actually effective in achieving flocculation. At high dose rates, coagulation obviously plays a part in improving settleability but the success reported with very low doses suggests a metabolic effect on the microbial populations present in the sludge. The exact nature of such processes has not been determined, but detailed studies of the effect of iron on activated sludge and filamentous organisms in activated sludge have been performed by Pfeffer *et al.* [5], Carter and McKinney [6], and Rensink *et al.* [7]. The last-named authors were able to control bulking by the continuous addition of 25 to 50 mg/l of iron to the influent sewage. They also noted that phosphorus removal occurred simultaneously.

Pipes [3] has mentioned the use of ferric salts and alum for bulking control.

A major disadvantage associated with addition of inorganic flocculants is increased sludge production. At high dose rates, large volumes of hydroxide sludge may be produced which can lead to subsequent sludge dewatering problems.

Thomanetz and Bardtke [8] were successful in controlling sludge bulking with synthetic polyelectrolyte flocculating agents. The very large range of such products currently available means that trials to find a suitable and effective coagulant can be very time-consuming.

COST OF CHEMICAL TREATMENT

Since chemical additons to control bulking are uncertain in their effectiveness, and because dose rates may vary from plant to plant, it is not possible to give a precise cost of this form of treatment. However, if certain assumptions are made, probable daily costs for dosing with various chemicals at works of different size can be derived. The following procedure was adopted:

(i) A 'typical' activated-sludge plant was considered. The plant was assumed to have a sewage retention period of six hours at Dry Weather Flow (DWF) and to carry a mixed-liquor suspended-solids concentration of 3000 mg/l. The latter figure may be high for a plant experiencing settlement problems but it allows for solids in the return sludge pipework and final tanks which are not accounted for separately.
An equivalence between chemical doses based on influent sewage flow (in mg/l) and dose based on mixed liquor suspended solids (kg chemical/kg MLSS) could then be established. Thus a dose rate of 30 mg/l into sewage is equivalent to MLSS dosing of 0.04 kg/kg MLSS.

(ii) Chemical costs were adjusted so that the price was established in terms of the 'active' substance in the chemical compound. Thus one tonne of ferrous sulphate in bulk costs about £20 and contains about 200 kg of iron. The cost of the iron in ferrous sulphate is therefore about £100/tonne. Similarly ferric sulphate is available in bulk as a concentrated solution containing about 40 per cent by weight of ferric sulphate. Purchased in this way, the iron in ferric sulphate costs about £600/tonne. It was necessary to establish costs on this basis because the dose rates quoted for each chemical are usually in terms of the 'active' substance or cation concentration.

(iii) Following this procedure it was possible to derive a simple formula for the daily cost of dosing with various chemicals at plants of different size. The costs for high and low dose rates are presented for convenience as nomographs in Figs. 1 and 2 respectively.

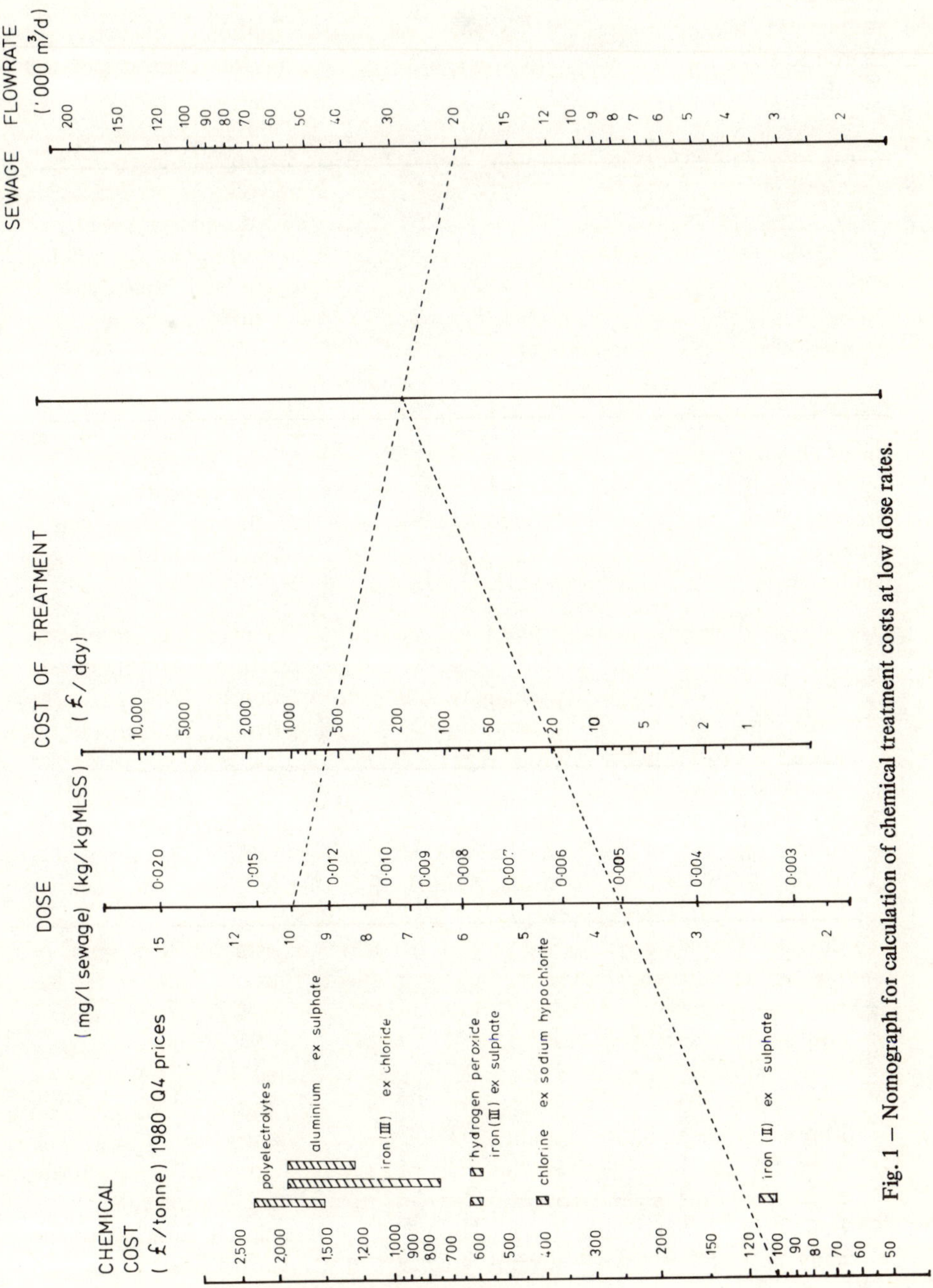

Fig. 1 — Nomograph for calculation of chemical treatment costs at low dose rates.

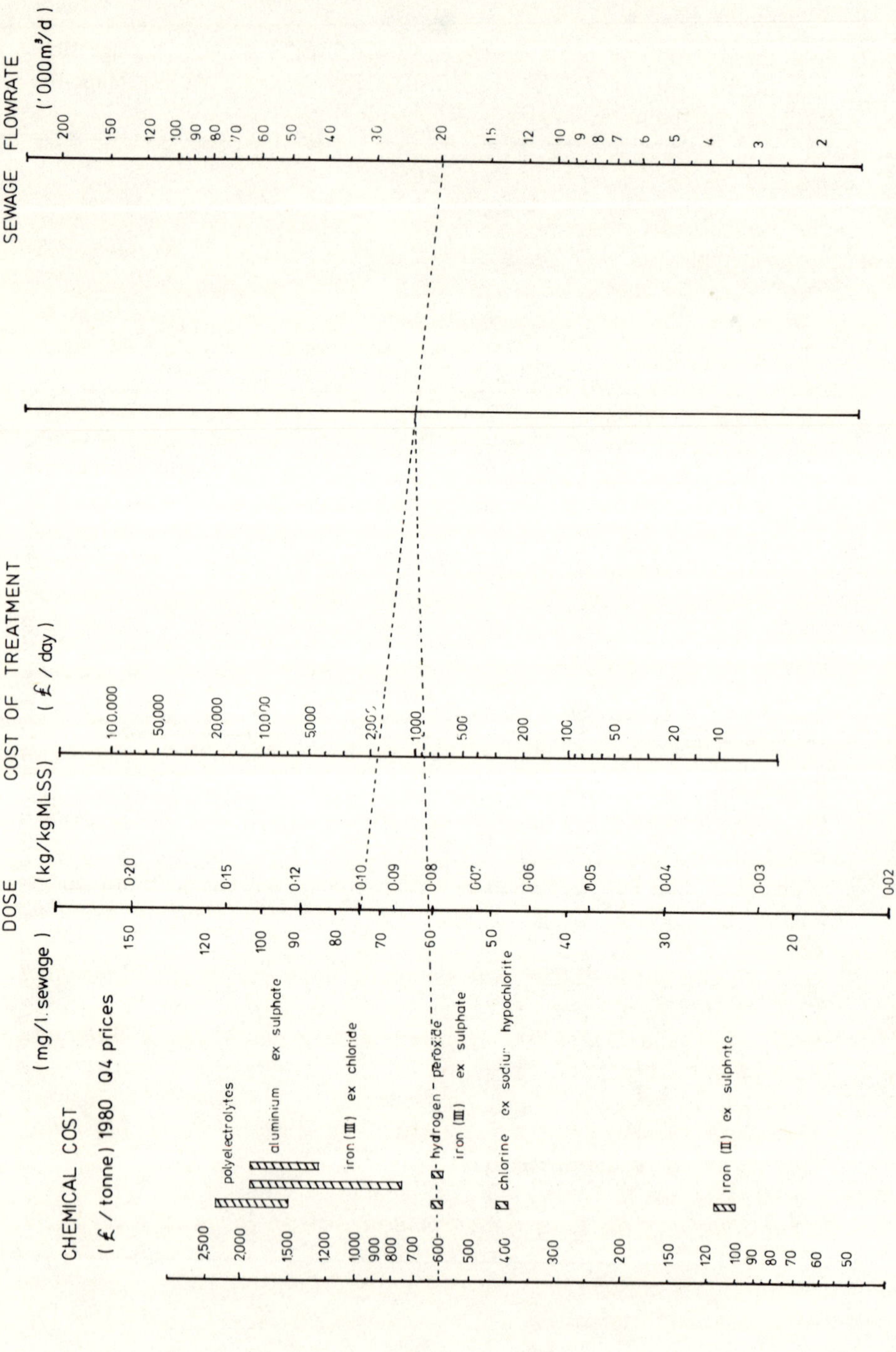

Fig. 2 – Nomograph for calculation of chemical treatment costs at high dose rates.

The nomographs are read as follows:

(i) Join the appropriate point on the right-hand scale, showing sewage flow rate, to the dose rate (either in mg/l of sewage or kg/kg MLSS). Mark the corresponding point on the ungraduated scale.

(ii) Join this point to the cost of the chosen chemical on the left-hand scale and read off the daily cost of treatment.

The probable cost ranges of various chemical additives are shown to the left of the nomographs. As stated previously these refer to the cost of the 'active' substance or cation and not the cost of the salt or solution containing the desired material. Chlorine costs are given, assuming sodium hypochlorite to be the chlorine-containing material. This is probably the simplest and least expensive method of adding chlorine to an activated-sludge plant.

The costs given in Figs. 1 and 2 do not include an allowance for the cost of the dosing equipment. On this basis the daily cost is seen to increase uniformly with plant throughput and chemical dose. Assuming the dose ranges given in the literature to be reasonable, then it can be readily seen that continuous dosing at high concentrations can be extremely costly in a large plant. Thus Keller and Cole's [4] stated minimum hydrogen peroxide dose of 0.1 kg/kg MLSS would cost over £900/day at a 20 000 m^3/d plant and nearly £5000/day at a 100 000 m^3/d plant.

REFERENCES

[1] Tapelshay, J. A. Control of sludge index by chlorination of return sludge. *Sewage Works Journal,* 1945, **17**, 1210–1226.

[2] Smith, R. S. and Purdy, W. C. Use of chlorination for correction of sludge bulking in activated sludge processes. *Sewage Works Journal,* 1936, **8**, 223–230.

[3] Pipes, W. O. Control bulking with chemicals. *Water and Wastes Engineering,* 1974, **11**, 30–34, 70–71.

[4] Keller, A. J. and Cole, C. A. Hydrogen peroxide cures bulking. *Water and Wastes Engineering,* 1973, **10**, E4, E6–7.

[5] Pfeffer, J. T., Samra, A. A. and Schwegler, D. T. Trace metals and filamentous micro-organism growth. *Proceedings 20th Industrial Waste Conference,* Purdue University, 1965, 608–617.

[6] Carter, J. L. and McKinney, R. E. Effects of iron on activated sludge treatment. *Journal of the Environmental Engineering Division, ASCE,* 1973, **99**, 135-152.

[7] Rensink, J. H., Leentvaar, J and Donker, H. J. Control of bulking combined with phosphate removal by ferrous sulphate dosing. *H_2O*, 1979, **12**, 150–153.

[8] Thomanetz, E. and Bardtke, D. Studies on the possibility for control of bulking sludge with synthetic flocculating agents. *Korrespondenz Abwasser*, 1977, **1824**, 15–17.

List of Delegates

ALEXANDER, A., Area Scientist (North), Thames Water Authority, Cotswold Division, 17 Bath Road, Swindon, Wiltshire, SN1 4AT.

ALLUM, K. H., Watson Hawksley, 201 Amersham Road, High Wycombe, Buckinghamshire.

ANDERSEN, K. K., Co-worker in Research Department, Akvadan-Harvey A/S, Krogshojvej 29, 2880 Bagsvaerd, Denmark.

ANDREWS, N. W., Sewage Works Manager (Oxford Area), Thames Water Authority, Vales Division, Oxford Sewage Works, Sandford on Thames, Oxfordshire. OX4 4YX.

ARUNDEL, J., Area Chemist, Thames Water Authority, Southern Division, Filmer Grove, Nightingale Road, Godalming, Surrey. GU7 3AD.

BARNES, L., District Manager, Anglian Water Authority, Norwich Sewage Division, PO Box 50, Thorpe Road, Norwich, Norfolk. NR1 1SA.

BLACHFORD, Miss A. J., WRC Process Engineering, Stevenage Laboratory, Elder Way, Stevenage, Hertfordshire. SG1 1TH.

BOON, A. G. Head, Process Development, Treatment Process Division, WRC Process Engineering, Stevenage Laboratory, Elder Way, Stevenage, Hertfordshire. SG1 1TH.

BOOTH, M. G., Regional Waste Treatment Scientist, Wessex Water Authority, Wessex House, Passage Street, Bristol, Avon.

BRANDSE, F. A., Zuivermgstechnoloog, Zuiveringschap West-Overyssel, Postbus 60, 8000 AB Zwolle, The Netherlands.

BRIGHT, S. G., Plant Chemist, Yorkshire Water Authority, Blackburn Meadows Sewage Works, Alsing Road, Sheffield. S9 1HF.

CHAMBARLHAC, Research Engineer, Degremont, 183 Avenue du 18 Juin 1940, 92 500 Rueil, France.

CHAMBERS, Dr. B., WRC Process Engineering, Stevenage Laboratory, Elder Way, Stevenage, Hertfordshire. SG1 1TH.

CHAPLIN, Mrs. M., Assistant Microbiologist, Welsh Water Authority, Meyler House, St Thomas Green, Haverfordwest, Dyfed. SA61 1QP.

CLAPHAM, G. S., Assistant Director of Operations, Wessex Water Authority, Passage Street, Bristol, Avon. BS2 0JQ.

CLARK, Mr. E. I., Principal Scientist, North West Water Authority, Dawson House, Great Sankey, Warrington, Cheshire. WA5 3LW.

CLOUGH, G. F. G., Consultant, Howard Humphreys and Partners, Thorncroft Manor, Leatherhead, Surrey.

COCKSEY, Dr. B. J., Northumbrian Water Authority, Wear Division Headquarters, Wear House, Abbey Road, Pity Me, Durham. DH1 5EZ.

COLE, S. R., Research Student, Department of Biology, University of York, Heslington, York. YO1 5DD.

COOPER. P. F., Section Leader, Wastewater Treatment Division, WRC Process Engineering, Stevenage Laboratory, Elder Way, Stevenage, Hertfordshire. SG1 1TH.

COTTON, P., Divisional Operations Controller, Anglian Water Authority, Norwich Sewage Division, PO Box 50, Thorpe Road, Norwich, Norfolk. NR1 1SA.

CRAIG, C. C., District Inspector, Forth River Purification Board, Colinton Dell House, Colinton, Edinburgh. EH13 0PH.

CRITCHARD, D. J., Area Chemist, Thames Water Authority, Lambourne Division, 24 Barossa Road, Camberley, Surrey.

CUTHBERT, Dr. J., Director, WRC Process Engineering, Stevenage Laboratory, Elder Way, Stevenage, Hertfordshire. SG1 1TH.

DAVIES, Dr. M., Senior Lecturer, Department of Biology, University of York, Heslington, York. YO1 5DD.

DENTON, R. S., Chemist, Whitehead and Poole Ltd, PO Box 9, Radcliffe, Manchester.

DEVENISH, I. W., Chief Chemist, Southern Water Authority, East Kent Water and Drainage Division, Westwood Road, Broadstairs, Kent.

EIKELBOOM, Ir. D. H., Research Institute for Environmental Hygiene, TNO, Postbus 214, Schoemakerstraat 9,2600 AE Delft, The Netherlands.

EVANCE, J. G., Area Manager, Anglian Water Authority, Essex Sewage Division, Braxted Park, Great Braxted, Near Witham, Essex.

EVERETT, K. M., PhD Student, Liverpool University, Department of Microbiology, Life Sciences Building, PO Box 147, Liverpool, Merseyside.

EVISON, L. M., Lecturer, University of Newcastle, Department of Civil Engineering, Claremont Road, Newcastle, Northumbria. NE1 7RU.

Van den EYNDE, E., University of Louvain, Laboratory of Industrial Microbiology and Biochemistry, Kard. Mercierlaan 92, Heverlee B3030, Belgium.

FORSTER, Dr. C. F., Lecturer, Department of Civil Engineering, University of Birmingham, PO Box 363, Birmingham. B15 2TT.

FRANKLIN, Ms. B. C., Yorkshire Water Authority, PO Box 201, Broadacre House, Vicar Lane, Bradford, West Yorkshire. BD1 5PZ.

FULLARD, J., Information Scientist, Unilever Research, Port Sunlight Laboratory, Quarry Road East, Bebington, Wirral, Merseyside.

GOODSON, M. G., Principal Engineer, Anglian Water Authority, Aqua House, London Road, Peterborough, Cambridgeshire.

GRAU, Professor, Dr. P., Head, Department of Water Technology and Environmental Engineering, 166 28 Prague 6, Czechoslovakia.

GRAY, Dr. N. F., Lecturer in Environmental Science, Trinity College, Environmental Sciences Unit, Dublin 2, Eire.

GREEN, Dr. G. P., Biologist, Regional Biology Unit, Wessex Water Authority, Avon and Dorset Division, 2 Nuffield Road, Poole, Dorset.

HALL, T., Research Assistant, Department of Biology, University of York, Heslington, York. YO1 5DD.

HEATH, D. W., Civil Engineering Manager, Wessex Water Authority, Wessex House, Passage Street, Bristol, Avon.

HEPWORTH, B. L., Lower Rother District Manager, Yorkshire Water Authority, Old Whittington Sewage Works, Newbridge Lane, Old Whittington, Chesterfield, Derbyshire.

HILTON, A. R., Assistant Controller Sewage Distribution, Southern Water Authority, East Sussex Water and Drainage Division, Construction House, Menzies Road, St Leonards on Sea, East Sussex.

HIRST, G., Leeds WPC Area Manager, Yorkshire Water Authority, North Central Division, 182 Otley Road, Leeds, Yorkshire. LS16 5PR.

HODGSON, I. R., Principal Engineer (Design), Wessex Water Authority, Bristol Avon Division, PO Box 95, Bath, Avon. BA1 2YP.

HORWOOD, C., Director, Ellis Horwood Ltd, Publishers, Market Cross House, Cooper Street, Chichester, West Sussex. PO19 1EB.

HORWOOD, E., Chairman, Ellis Horwood Ltd, Publishers, Market Cross House, Cooper Street, Chichester, West Sussex. PO19 1EB.

HOUSTON, K. C., Area Controller Sewage and Drainage, Welsh Water Authority, Usk Division, Station Buildings, Queensway, Newport, Gwent. NP4 4AX.

HOUTMEYERS, J., Forest Biology Centre, Bokrijk-Genk, Belgium.

HUDSON, Dr. J. A., Harrogate WPC Area Manager, Yorkshire Water Authority, North Central Division, 182 Otley Road, Leeds, Yorkshire. LS16 5PR.

HULME, Dr. P., Principal Scientific Officer (Treatment), Welsh Water Authority, Cambrian Way, Brecon, Powys. LO3 7RR.

HURLEY, B. J. E., Principal Research Scientist, Thames Water Authority, Directorate of Scientific Services, New River Head, Rosebery Avenue, London. EC1R 4TP.

HUTTON, J. L., Chief Assistant, Grampian Regional Council, Department of Water Services, Woodhill House, Ashgrove Road West, Aberdeen. AB9 2LU.

JENKINS, Professor D., Department of Sanitary Engineering, University of California, Berkeley, 1301 South 46th Street, Richmond, California 94804, USA.

JOLLIFFE, R. F., Senior Scientific Assistant, Severn-Trent Water Authority,

Avon Division, Avon House, De Montford Way, Coventry, Midlands.

JOHNSON, G. R., Sales Manager Wimpey Unox Ltd, Gunnersbury House, 626 Chiswick High Road, London. W4 5RZ.

JOHNSTONE, Dr. D. W. M., Divisional Scientist, Thames Water Authority, Cotswold Division, 17 Bath Road, Swindon, Wiltshire. SN1 4AT.

KEEGAN, I. D., Assistant Works Chemist, British Sugar Corporation, PO Box 17, Boroughbridge Road, York. YO1 1XF.

KELLOCK, R. W., Group Manager, Thames Water Authority, East Hyde Sewage Treatment Works, Lower Luton Road, Near Luton, Bedfordshire.

KIMBER, A., Wessex Water Authority, Avon and Dorset Division, 2 Nuffield Road, Poole, Dorset. BH17 7RL.

KING, R. P., Divisional Scientist, Thames Water Authority, Metropolitan Public Health Division, 50–64 Broadway, London. SW1H 0DB.

KIRBY, R. M., Area Chemist, Anglian Water Authority, Essex Sewage Division, Braxted Park, Great Braxted, Near Witham, Essex.

KITE, O. A., Design Engineer, BOC Limited, Deer Park Road, London. SW19 3UF.

LATTEN, A., Works Manager, Anglian Water Authority, Norwich Sewage Division, PO Box 50, Thorpe Road, Norwich, Norfolk. NR1 1SA.

LAU, A. O., Research Engineer, EXXON Research and Engineering, PO Box 101, Florham Park, New Jersey 07960, USA.

McBREARTY, G., Area Chemist, Thames Water Authority, Southern Division, Filmer Grove, Nightingale Road, Godalming, Surrey. GU7 3AD.

McFARLANE, D., Structural Graduate Engineer, Ove Arup and Partners, 13 Fitzroy Street, London. W1P 6BQ.

McQUEEN, A., Trade Effluent Officer, Tayside Regional Council, Water Services Department, Clatto Treatment Works, Balmahon Drive, Dundee. DD3 9RP.

MA, A. N., Research Officer, Palm Oil Research Institute of Malaysia (Porim), 18th Floor, Angkasa Raya Building, Jalan Ampang, Kuala Campur, Malaysia.

MARPOLE, D. W., Group Manager, Thames Water Authority, Lea Division, The Grange, Crossbrook Street, Cheshunt, Hertfordshire.

MATCHETT, R. J. H., Head of Civil Engineering (Sewage Treatment), Thames Water Authority, Metropolitan Public Health Division, 50–64 Broadway, London. SW1H 0DB.

MAXWELL, P. J., Principal Operation Controller (Reclamation), Severn-Trent Water Authority, Lower Severn Division, Southwick Park, Gloucester Road, Tewkesbury, Gloucestershire.

MORRISS, A. C., Thames Water Authority, Reading Sewage Treatment Works, Manor Farm Road, Reading, Berkshire.

NEWTON, G., Assistant Area Controller, Severn-Trent Water Authority, Lower Trent Division, Mapperley Hall, Lucknow Avenue, Mapperley, Nottingham. NG3 5BN.

NEWMAN, Dr. P. J., Manager, European Division, WRC Environmental Protection, Medmenham Laboratory, PO Box 16, Henley Road, Medmenham, Marlow, Buckinghamshire. SL7 2HD.

NOLON, R. P., Senior Chemist, Treatment Division, Wessex Water Authority, PO Box 95, Bath, Avon. BA1 2YP.

O'NEILL, J., Division WPC Manager, Yorkshire Water Authority, Western Division, PO Box 201, Broadacre House, Vicar Lane, Bradford, West Yorkshire. BD1 5PZ.

OSMOND, P. H., Area Water Pollution Control Manager, Southern Water Authority, Westwood Road, Broadstairs, Kent. CT10 2NZ.

PIKE, Dr. E. B. Head of Microbiology, WRC Process Engineering, Stevenage Laboratory, Elder Way, Stevenage, Hertfordshire. SG1 1TH.

POTTEN, A. H., Scientific Officer, Anglian Water Authority, Essex Sewage Division, Braxted Park, Great Braxted, Witham, Essex.

PREEDY, Mrs. E. A., Senior Engineer, John Taylor and Sons, Artillery House, Artillery Row, London. SW1P 1RY.

PREEDY, J. J., Project Manager, Hawker Siddeley Water Engineering Ltd, Molly Millars Lane, Wokingham, Berkshire.

PUJOL, R., Ingenieur, Ministerie de l'Agriculture – CEMAGREF, 11a Avenue de Saint Maude, Paris 75012, France.

RACHWAL, A. J., Research Scientist, Thames Water Authority, Central Research Laboratory, Northumberland House, Mogden Sewage Treatment Works, Mogden Lane, Isleworth, Middlesex.

RENNERFELT, L., President, Ferriklor International AB, Fredrikshovsgat 3A, S-11522 Stockholm, Sweden.

RENSINK, Ir. J. H., Department of Water Pollution Control, Agricultural University, Biotechnion, De Dreijen 12/6703 BC, Wageningen, The Netherlands.

RICHARDS, S., Research Student, Department of Biology, University of York, Heslington, York. YO1 5DD.

RIMMER, T. T., Plant Chemist, Yorkshire Water Authority, Blackburn Meadows Sewage Works, Alsing Road, Sheffield. S9 1HF.

ROBERTS, D. H. L., Principal Water Reclamation Officer, Severn-Trent Water Authority, Soar Division, Leicester Water Centre, Gorse Hill, Anstey, Leicestershire.

ROSE, R. L., Process Engineer, Wimpey Unox Ltd, Gunnersbury House, 626 Chiswick High Road, London. W4 5RZ.

ROWELL, M. R., Principal Scientist, Severn-Trent Water Authority, Lower Severn Division, Southwick Park, Gloucester Road, Tewkesbury, Gloucestershire.

RUNDLE, H., Principal Scientist, Severn-Trent Water Authority, Lower Trent Division, Mapperley Hall, Lucknow Avenue, Mapperley, Nottingham. NG3 5BN.

SALANITRO, J. P., Microbiologist, Shell Development Company, West Hollow

Research Centre, PO Box 1380, Houston, Texas, 77001, USA.

SCHULZ, H., General Manager, Ferriklor International (Deutschland GmbH) Heinr Fendelstr 18, D-6240 Limburg, West Germany.

SLIJKHUIS, H., Department of Microbiology, Agricultural University, Hesselink van Suchtelenweg 4/6703, Wageningen, The Netherlands.

STEPHENSON, R. J., Department of Fermentation and Microbiology, Shell Research Ltd, Sittingbourne, Kent.

STOUNAUG, R., Research Sanitary Engineer, NTNF's Committee for Treatment Plant Operation Research, PO Box 333, Blindern, Oslo 3, Norway.

Van den STRATEN, J., Waterboard de Dommel, Molenpad 8, Boxtel, The Netherlands.

TAYLOR, J. M., Division WPC Manager, Yorkshire Water Authority, Copley House, Waterdale, Doncaster, Yorkshire. DN1 3JF.

THOMPSON, Dr. K. C., Principal Scientific Officer, Yorkshire Water Authority, Blackburn Meadows Sewage Works, Alsing Road, Sheffield. S9 1HF.

TOMLINSON, E. J., WRC Process Engineering, Stevenage Laboratory, Elder Way, Stevenage, Hertfordshire. SG1 1TH.

TRAMONTINI, Mrs. E., Wessex Water Authority, Avon and Dorset Division, 2 Nuffield Road, Poole, Dorset.

VARLEY, R. A., Special Projects Officer, Wessex Water Authority, Avon and Dorset Division, 2 Nuffield Road, Poole, Dorset. BH17 7RL.

VERACHTERT, Professor H., University of Louvain, Laboratory of Industrial Microbiology and Biochemistry, Kard. Mercierlaan 92, Heverlee B3030, Belgium.

VRIES, de F. A., Principal (Operations), Southern Water Authority, Guildbourne House, Chatsworth Road, Worthing, Sussex.

WAGNER, F., Institute of Water Conservation, Water Purification and Industrial Waste, University of Stuttgart, Bandtale 1, 7000 Stuttgart 80, West Germany.

WAKEFIELD, M., Divisional Scientific Officer, Wessex Water Authority, Somerset Division, PO Box 9, Kings Square, Bridgewater, Somerset.

WAITE, W. M., Principal Microbiologist, Severn-Trent Water Authority, Abelson House, 2297 Coventry Road, Birmingham.

WALLER, C. B., Divisional Scientist, Thames Water Authority, Vales Division, Denton House, Iffley Turn, Oxford.

WATKINS, J., Division Bacteriologist, Yorkshire Water Authority, Copely House, Waterdale, Doncaster, Yorkshire. DN1 3JF.

WEBB, L. J., Group Head, PIRA, Randalls Road, Leatherhead, Surrey. KT22 7RU.

WHEALE, G. A. WPC Chemist (Operations), Yorkshire Water Authority, Thrum Hall Lane, Halifax, Yorkshire.

WHITE, Dr. M. J. D., Assistant Director, WRC Process Engineering, Stevenage Laboratory, Elder Way, Stevenage, Hertfordshire. SG1 1TH.

WICKS, W. R., Severn-Trent Water Authority, Leicester Water Centre, Gorse Hill, Anstey, Leicestershire.

WILLIAMS, D. A., Section Manager, Unilever Research Ltd, Port Sunlight Laboratory, Quarry Road East, Bebington, Wirral, Merseyside.

WOUDA, T., Waterboard de Dommel, Molenpad 8, Boxtel, The Netherlands.

WRIGHT, J. F., Inspecting Engineer, Welsh Water Authority, Cambrian Way, Brecon, Powys.

Index